$$2x - 7 = -13$$
$$+ 7 = +7$$
$$\frac{2x}{2} = \frac{-6}{2}$$
$$x = -3$$

BARRON'S
E-Z
PRE-ALGEBRA

Caryl Lorandini

BARRON'S

Better Grades or Your Money Back!

As a leader in educational publishing, Barron's has helped millions of students reach their academic goals. Our E-Z series of books is designed to help students master a variety of subjects. We are so confident that completing all the review material and exercises in this book will help you, that if your grades don't improve within 30 days, we will give you a full refund.

To qualify for a refund, simply return the book within 90 days of purchase and include your store receipt. Refunds will not include sales tax or postage. Offer available only to U.S. residents. Void where prohibited. Send books to **Barron's Educational Series, Inc., Attn: Customer Service** at the address on this page.

All inquiries should be addressed to:
Barron's Educational Series, Inc.
250 Wireless Boulevard
Hauppauge, New York 11788
www.barronseduc.com

ISBN: 978-1-4380-0011-4

Library of Congress Catalog Card No. 2012010558

Library of Congress Cataloging-in-Publication Data
Lorandini, Caryl.
    Barron's E-Z pre-algebra / Caryl Lorandini.
       p.   cm.
    Summary: "A thorough preparation for students who need the basic skills to continue on to Algebra classes, and an accessible review for older students who need to brush up on forgotten rules and techniques"—Provided by publisher.
    Includes index.
    ISBN 978-1-4380-0011-4   (pbk.)
    1. Algebra—Problems, exercises, etc.   I. Title.
II. Title: E-Z pre-algebra.
QA157.L67 2012
512.9—dc23

                                                                    2012010558

PRINTED IN THE UNITED STATES OF AMERICA
9 8 7 6 5 4 3 2 1

CONTENTS

Introduction — v

1 Properties and Order of Operations — 1
1.1 Numbers and Sets — 1
1.2 Properties of Real Numbers — 4
1.3 Order of Operations — 5
Chapter Review — 7

2 Fractions — 9
2.1 Equivalent Fractions — 10
2.2 Adding Fractions — 12
2.3 Subtracting Fractions — 15
2.4 Multiplying Fractions — 18
2.5 Dividing Fractions — 19
2.6 Word Problems — 20
Chapter Review — 22

3 Decimals — 23
3.1 How Do We Read, Write, and Compare Decimals? — 23
3.2 How Do We Add and Subtract Decimals? — 26
3.3 How Do We Multiply Decimals? — 28
3.4 How Do We Divide Decimals? — 30
Chapter Review — 31

4 Integers — 33
4.1 What Are Integers? — 33
4.2 Adding Integers — 35
4.3 Subtracting Integers — 37
4.4 Multiplying and Dividing Integers — 38
Chapter Review — 42

5 Variables and Expressions — 43
5.1 Using Symbols — 44
5.2 Writing Expressions — 45
5.3 Simplifying Expressions — 46
5.4 Evaluating Expressions — 48
Chapter Review — 49

6 Ratios and Proportions — 51
6.1 What Is a Ratio? — 52
6.2 What Is a Rate? — 54
6.3 Comparing Unit Rates — 55
6.4 What Is a Proportion? — 55
6.5 Solving Proportions — 57
6.6 Similar Figures — 58
6.7 Scale Drawings — 60
6.8 Word Problems — 62
Chapter Review — 64

7 Percents and Percentages — 65
7.1 What Is a Percent? — 66
7.2 How Do We Calculate a Percent? — 67
7.3 How Do We Find the Whole, Given the Percent? — 70
7.4 How Do We Find the Percent? — 71
7.5 When Is Percent Used? — 72
7.6 What Is Percent Increase or Decrease? — 78
7.7 What Is Percent Error? — 80
Chapter Review — 81

8 Factors and Exponents — 83
8.1 What Are Factors, Factor Pairs, and Exponents? — 84
8.2 What Is Prime Factorization? — 87
8.3 What Are Common Factors and Multiples? — 89
8.4 What Are the Laws of Exponents? — 91
8.5 What Is Scientific Notation? — 94
8.6 How Can We Evaluate Expressions with Exponents? — 96
Chapter Review — 97

9 Solving Equations — 99
9.1 How Do We Solve a One-Step Equation? — 100
9.2 How Do We Solve a Two-Step Equation? — 102
9.3 How Can Combining Like Terms Help Us? — 104
9.4 How Can the Distributive Property Help Us? — 106
9.5 How Can We Solve Multi-Step Equations? — 107
9.6 How Can We Write an Equation to Solve Problems Algebraically? — 109
Cumulative Review — 111

10 Solving Inequalities — 113
10.1 What Is an Inequality? — 114
10.2 How Do We Represent Solutions of Inequalities? — 115
10.3 How Do We Solve Inequalities? — 117
10.4 How Can We Use Inequalities to Solve Word Problems? — 119
Chapter Review — 121

11 Geometry   123
11.1 What Are Angle Pair Relationships?   124
11.2 What Are Vertical Angles?   127
11.3 What Happens When Parallel Lines   130
Are Cut by a Transversal?
11.4 How Do We Classify Two-Dimensional   135
Figures?
11.5 What Is Area and Perimeter and   138
How Do We Calculate Them?
11.6 How Do We Classify Three-Dimensional   143
Figures?
11.7 What Is Surface Area and Volume   146
and How Do We Calculate Them?
Chapter Review   149

12 Functions and Graphing   151
12.1 What Is Coordinate Geometry?   153
12.2 How Do We Graph a Line from a Table   155
of Values?
12.3 What Is Slope?   158
12.4 How Do We Graph a Line Using Slope?   160
12.5 What Is a Function?   164
12.6 What Is the Rule of Four?   168
12.7 How Can We Tell If It Is Linear   171
or Non-linear?
Chapter Review   174

13 Transformational Geometry   175
13.1 What Is Transformational Geometry?   175
13.2 What Is a Reflection?   179
13.3 What Is a Translation?   181
13.4 What Is Rotation?   185
13.5 What Is a Dilation?   188
13.6 What Is Symmetry?   191
Chapter Review   193

14 System of Linear Equations   195
14.1 What Are the Possibilities When Solving   195
Simultaneous Linear Equations?
14.2 How Do We Solve a System Graphically?   197
14.3 How Do We Solve a System by Addition?   200
14.4 How Do We Solve a System by Substitution?   202
14.5 How Do We Solve Word Problems   204
Leading to Two Linear Equations?
Chapter Review   207

15 The Pythagorean Theorem   209
15.1 What Is the Pythagorean Theorem?   210
15.2 What Is the Converse of the   211
Pythagorean Theorem?
15.3 How Can We Use the   212
Pythagorean Theorem?
15.4 How Can We Use the Pythagorean   215
Theorem on the Coordinate Plane?
Chapter Review   216

16 Probability   219
16.1 What Are Tree Diagrams and the   220
Counting Principle?
16.2 What Is Probability?   223
16.3 How Do We Find the Probability of   224
Simple Events?
16.4 How Do We Find the Probability of   226
Compound Events?
16.5 What Is a Permutation?   229
16.6 What Is a Combination?   230
Chapter Review   232

17 Statistics   233
17.1 What Are Statistics?   234
17.2 What Are Measures of Central Tendency?   240
17.3 How Do We Interpret Data?   242
17.4 How Do We Create Dot Plots?   246
17.5 How Do We Create a Stem-Leaf Plot?   248
17.6 How Do We Create a Box-Whisker?   250
17.7 How Do We Create a Histogram?   252
17.8 How Do We Create a Scatter Plot?   255
Chapter Review   260

18 Sequences   263
18.1 What Are Patterns?   263
18.2 What Are Arithmetic Sequences?   265
18.3 What Are Geometric Sequences?   266
Chapter Review   268

Cumulative Review   269
Answers to Chapter Reviews   275
Index   287

INTRODUCTION

Everyone should be able to see mathematics as an exciting and useful means of communication. We express our thoughts and ideas to each other using shortcuts and symbols. We all understand the text language of "lol" or "brb." The language of mathematics isn't very different from text language. The more we use the language, the simpler the symbols or shortcuts become to us. The intent of this book is to solidify the concepts and basic understanding needed for success in algebra. But what is algebra? Algebra is the branch of mathematics that uses letters, symbols, and characters to represent numbers and express mathematical relationships. The better we build our foundation, the bigger and more expansive our mathematical knowledge can be built upon it. Algebra and geometry concepts are interconnected; it is this interconnectedness that strenghthens substantive meaningful learning.

Properties and Order of Operations

WHAT YOU WILL LEARN

- Some basic terms and assumptions of numbers and sets
- The difference between rational and irrational numbers
- Properties of real numbers
- Order of operations

SECTIONS IN THIS CHAPTER

- Numbers and Sets
- Properties of Real Numbers
- Order of Operations

1.1 Numbers and Sets

In our daily lives we use numbers to count, compare, order, measure, and solve problems. **Numbers** are the concept of the amount or quantity in a collection. The numbers we use the most and are most familiar with are whole numbers, but let's look at some different sets of numbers.

STARTING SMALL AND BUILDING UP

Counting numbers   All whole numbers greater than zero; also called natural numbers: {1,2,3,4,...}.

Whole numbers   The set of counting numbers plus zero: {0,1,2,3,...}.

Integers   The set of numbers consisting of the counting numbers (1, 2, 3, 4, ...), their opposites (−1, −2, −3, −4, ...), and zero.

Rational numbers   Any number that can be expressed as a ratio in the form $\frac{a}{b}$ where a and b are integers and $b \neq 0$. Rational numbers include all counting, whole, and integer numbers. They also include all fractions and most decimal numbers.

Decimal number   A fractional number written using base ten notation; a mixed decimal number has a whole number part as well (e.g., 0.32 is a decimal number and 3.5 is a mixed decimal number).

Fraction   A number that represents part of a whole, part of a set, or a quotient in the form $\frac{a}{b}$ which can be read as a divided by b.

Irrational number   A real number that cannot be represented as an exact ratio of two integers; the decimal form of the number never terminates and never repeats (e.g., π, $\sqrt{2}$, $\sqrt{10}$, 0.010010001...).

Real numbers   The set of numbers that includes all rational and irrational numbers.

THINGS TO THINK ABOUT:
Can decimals be turned into fractions?

When you read the decimal using place values, you are actually doing that.

.8 can be read as "eight-tenths" or 8/10; therefore, it is rational.

1.8674 can be read as "one and eight thousand six hundred seventy-four ten-thousandths" or 1 and 8674/10,000; therefore, it is rational.

.139 can be read as "one hundred thirty-nine thousandths" or 139/1000; therefore, it is rational.

What about square roots?

$\sqrt{16}$ When calculated, this has a principal root of 4; therefore, it is rational.

$\sqrt{70}$ When calculated, this has a decimal form of a number that never terminates and never repeats; therefore, it is irrational.

Real Numbers

Rational   Irrational

integers

whole

counting
numbers

**EXAMPLE
1.1**

1) Name all of the sets to which the following numbers belong:
   a) 0
   b) −6
   c) $\frac{1}{2}$
   d) π
   e) 2.3

2) Tell whether rational or irrational.
   a) $\frac{2}{6}$
   b) $\sqrt{25}$
   c) $\frac{3}{11}$
   d) .010010001
   e) $\sqrt{35}$
   f) 2.33

SOLUTIONS

1) a) whole, integers, rational, real
   b) integers, rational, real
   c) rational, real
   d) irrational, real
   e) rational, real

2) a) rational
   b) rational
   c) rational
   d) irrational
   e) irrational
   f) rational

1.2 Properties of Real Numbers

From our experiences with numbers we notice patterns and have understandings. Many of these are the Properties of Real Numbers. Most of us just use these properties because we are so familiar with them that we don't even realize they are facts or truths that once had been proven—sort of like old friends that have always been there.

Commutative property   A property of real numbers that states that the sum or product of two terms is unaffected by the order in which the terms are added or multiplied.

The sum remains the same.

$$2 + 3.5 = 3.5 + 2 \qquad \tfrac{1}{2} + 7 = 7 + \tfrac{1}{2}$$

The product remains the same.

$$3 \times 5 = 5 \times 3 \qquad 5 \cdot x = x \cdot 5$$

Associative property   A property of real numbers that states that the sum or product of a set of numbers is the same, regardless of how the numbers are grouped.

The sum remains the same.

$$2 + (3.5 + 1.3) = (2 + 3.5) + 1.3$$

The product remains the same.

$$6 \times (18 \times 7) = (6 \times 18) \times 7$$

Additive identity   The number in a set which, when added to any number n in the set, yields the given number; the identity element for addition is zero because $n + 0 = n$ and $0 + n = n$. This is sometimes called the **addition property of zero**.

Multiplicative identity   The number in a set which, when multiplied by any number n in the set, yields the given number; the identity element for multiplication is one because $n \times 1 = n$ for all n in the set. This is sometimes called the **multiplicative property of one**.

Multiplicative property of zero   Any number that when multiplied by zero gives the product of zero.

$$8 \times 0 = 0$$

Additive inverse   A number that, when added to a given number, results in a sum of zero; the opposite of a number.

$$5 + -5 = 0$$

Remember: zero is the additive identity.

Multiplicative inverse   The reciprocal of a number; a number that, when multiplied by a given number, results in a product of 1.

$$7 \times \tfrac{1}{7} = 1 \text{ and } \tfrac{2}{3} \times \tfrac{3}{2} = 1$$

Remember: one is the multiplicative identity.

Distributive property   A property of real numbers that states that the product of the sum or difference of two numbers is the same as the sum or difference of their products.

Multiplication over addition:

$$2(15 + 4) = 2 \times 15 + 2 \times 4$$

Multiplication over subtraction:

$$4(12 - 8) = 4 \times 12 - 4 \times 8$$

EXAMPLE 1.2

Tell whether each statement is true or false
a) $4 + 3 = 3 + 4$
b) $10 \times 4 = 4 \times 10$
c) $6 - 8 = 8 - 6$
d) $6 \times 0 = 0 \times 9$
e) $35 + 0 = 35 \times 1$
f) $13 + 0 = 13 \times 0$

SOLUTION
a) True
b) True
c) False
d) True
e) True
f) False

1.3 Order of Operations

Most of us are not fans of rules, but sometimes rules are necessary. Rules will help us to avoid confusion when calculating; they will make sure we all know the same order in which to carry out operations. These rules allow everyone everywhere to get the same answers to calculations.

Exponent   A number that tells how many times the base is used as a factor; in an expression of the form b^a, a is called the exponent, b is the base, and b^a is a power of b.

Square a number To multiply a number by itself (e.g., $4 \times 4 = 16$ or $4^2 = 16$).

Square root of a number A number (factor) that when multiplied by itself yields the original number (e.g., $\sqrt{16} = 4$ or $\sqrt{16} = -4$).

ORDER IS IMPORTANT

To evaluate numerical expressions with more than one operation, we use order of operations.

1) Parentheses—all calculations within the parentheses, including braces, above or below a division bar, and absolute value.

2) Powers—all exponents and roots (for more information, see Chapter 8)

3) Multiplication or division

   • In order from left to right

4) Addition or subtraction

   • In order from left to right

Examples:

$9 - 3 \times 2 + 7$	do the multiplication
$9 - 6 + 7$	do the subtraction
$3 + 7$	do the addition
10	

$12 \div 3 + 5 \times 8$	do the division
$4 + 5 \times 8$	do the multiplication
$4 + 40$	do the addition
44	

$2^4 + 32 \div 8 - 6$	do the power ($2^4 = 2 \times 2 \times 2 \times 2$)
$16 + 32 \div 8 - 6$	do the division
$16 + 4 - 6$	do the addition
$20 - 6$	do the subtraction
14	

$40 \div 2 + (5 - \sqrt{4})$	do the operation in the parentheses (root)
$40 \div 2 + (5 - 2)$	do the operation in the parentheses (subtract)
$40 \div 2 + 3$	do the division
$20 + 3$	do the addition
23	

EXAMPLE 1.3

a) $3 \times 2 + 7 \times 3$

b) $18 \div 3 \times 2$

c) $\dfrac{18 + 2 \times 3}{5 - 2}$

d) $4^2 - (6 + 2) \div 2$

e) $31 + \sqrt{9} - 16 \div 4$

SOLUTION

a) 27

b) 12

c) 8

d) 12

e) 30

Chapter Review

1) Each number in the first column belongs to at least one of the sets of numbers displayed in the top row. Check ALL that are true.

	Counting Number	Whole Number	Integer	Rational	Irrational	Real
5	C N					
0	C N					
−2	I					
$\frac{1}{2}$						
π		I R				
$\sqrt{6}$		I R				
$\sqrt{16}$		I R				

2) Match the example to the property.

1. $9 \times 3 = 3 \times 9$
2. $15 + 0 = 15$
3. $2(3 - 6) = 6 - 12$
4. $3 \times (4 \times 5) = (3 \times 4) \times 5$
5. $4 \times \frac{1}{4} = 1$
6. $6 + 8 = 8 + 6$
7. $4 \times 0 = 0$
8. $3 + -3 = 0$

a) Commutative property of addition
b) Commutative property of multiplication
c) Associative property of multiplication
d) Additive identity
e) Multiplication by Zero
f) Additive inverse
g) Multiplicative inverse
h) Distributive property

3) Simplify

$$2^2 + 9 - (12 - 4)$$

$$\frac{10 + 3 - 7}{4}$$

$$8 \times (2 - 5)^2 \times 3$$

4) Check the work shown. Is the answer correct or incorrect? If incorrect, show the correct way to complete the problem.

$$50 - 18 \div 2 \times 3$$
$$50 - 18 \div 6$$
$$50 - 3$$
$$47$$

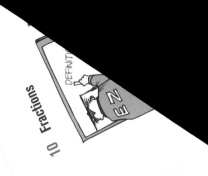

Fractions

WHAT YOU WILL LEARN

- The equivalence of fractions in special cases, and how to compare fractions by reasoning about their size.
- How to express a fraction as an equivalent fraction with a different denominator.
- How to decompose fractions to justify their sum or difference.
- How to add and subtract fractions with unlike denominators (including mixed numbers).
- How to interpret a fraction as a division of the numerator by the denominator ($\frac{a}{b} = a \div b$). Solve word problems involving division of whole numbers, leading to answers in the form of fractions or mixed numbers.
- How to apply and extend previous understandings of multiplication to multiply a fraction or whole number by a fraction.
- How to interpret and compute quotients of fractions, and solve word problems involving division of fractions.

SECTIONS IN THIS CHAPTER

- Equivalent Fractions
- Adding Fractions
- Subtracting Fractions
- Multiplying Fractions
- Dividing Fractions
- Word Problems

Fraction A number that represents part of a whole, part of a set, or a quotient in the form $\frac{a}{b}$, which can be read as *a* divided by *b*.

Denominator The quantity below the line in a fraction. It represents the number of equal parts into which the whole is divided.

Equivalent fractions Two or more fractions that have the same quotient or that name the same region, part of a set, or part of a segment.

Example: $\frac{1}{3} = \frac{3}{9}$

Improper fraction A fraction whose numerator is greater than its denominator.

Lowest common denominator (LCD) The smallest common multiple of two given denominators (e.g., the LCD of $\frac{1}{3}$ and $\frac{1}{8}$ is 24).

Least common multiple (LCM) The LCM of two numbers is the smallest number (that is not zero) that is a multiple of both.

Mixed number A number composed of an integer and a proper fraction (e.g., $3\frac{2}{9}$).

Numerator The top number in a fraction; it tells the number of equal parts (numerator) out of the total number of parts (denominator) being described by the fraction.

Proper fraction A fraction whose numerator is less than its denominator.

Reciprocal The number that is used to multiply a given number to obtain an answer of 1.

Unit fraction A fraction with a numerator of 1.

Unlike denominators Two or more fractions with unequal denominators.

2.1 Equivalent Fractions

A fraction describes a part of something. It can be part of a group or part of a whole. There is a bottom part, called the **denominator**, which represents the number of parts in the whole. There is the top part, called the **numerator**, which tells you how many of those parts you actually have.

Here are some examples:

$\frac{3}{4}$

$\frac{5}{6}$

There are many ways to tell if fractions are equivalent. The simplest way is to simplify the fraction to lowest terms, meaning there are no more common factors between the numerator and denominator.

Example:

$$\frac{2}{3} \text{ and } \frac{14}{21}$$

$\frac{2}{3}$ is already in lowest terms

$\frac{14}{21}$ has a common factor of 7 between the numerator and denominator. Divide them both by 7 and you get the simplified fraction of $\frac{2}{3}$. $\frac{2}{3}$ is the same as $\frac{2}{3}$, therefore $\frac{2}{3} = \frac{14}{21}$.

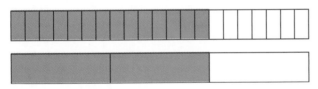

$$\frac{15}{45} \text{ and } \frac{6}{18}$$

$\frac{15}{45}$ has a common factor of 15 between the numerator and denominator. Divide them both by 15 and you get the simplified fraction of $\frac{1}{3}$.

$\frac{6}{18}$ has a common factor of 6 between the numerator and denominator. Divide them both by 6 and you get the simplified fraction of $\frac{1}{3}$. $\frac{1}{3}$ is the same as $\frac{1}{3}$, therefore $\frac{15}{45} = \frac{6}{18}$.

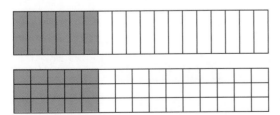

NOTE: When you divide or multiply the numerator AND denominator by the same number, it is equal to dividing or multiplying by 1. Since 1 is the multiplicative identity, it does not change the value of the fraction.

This can be useful when determining how many items are needed in a group. You want $\frac{2}{3}$ of the marbles in your vase to be blue. The vase can hold 36 marbles.

$\frac{2}{3}$ needs to be rewritten with a denominator of 36.

In the prior examples we divided the denominator and numerator by the same number to reduce the fraction. In this case we want the denominator and numerator to be larger, so we will multiply. Since 3 times 12 is 36, 12 is the number we need.

$$2 \times 12 = 24 \qquad 3 \times 12 = 36 \qquad \frac{2}{3} = \frac{24}{36}$$

You would want 24 blue marbles.

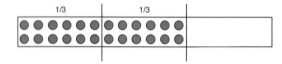

EXAMPLE 2.1

Tell whether the fractions are equivalent.

1) $\frac{12}{24}$ and $\frac{3}{6}$

2) $\frac{6}{9}$ and $\frac{2}{3}$

3) $\frac{5}{6}$ and $\frac{25}{36}$

4) $\frac{8}{9}$ and $\frac{10}{11}$

5) $\frac{16}{34}$ and $\frac{40}{85}$

SOLUTION

1) Yes

2) Yes

3) No

4) No

5) Yes

2.2 Adding Fractions

Adding fractions is a matter of counting. If you have $\frac{2}{9}$ and your friend has $\frac{4}{9}$, together you have $\frac{6}{9}$ or $\frac{2}{3}$.

Remember, $\frac{2}{9}$ is just $\frac{1}{9}$ and $\frac{1}{9}$ added together to make $\frac{2}{9}$. So you have $\frac{1}{9} + \frac{1}{9}$ and your friend has $\frac{1}{9} + \frac{1}{9} + \frac{1}{9} + \frac{1}{9}$, together you have $\frac{6}{9}$.

This is rather simple when you have the same sized pieces or the same denominator, because you are just counting up your pieces.

$$\frac{5}{12} + \frac{4}{12}$$

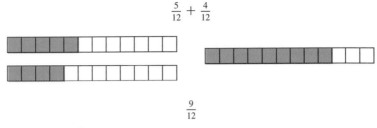

$$\frac{9}{12}$$

Can be simplified to $\frac{3}{4}$

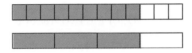

When fractions are being added together, you need to have like denominators. Like denominators come from the LCM of the denominators, called the Lowest Common Denominator (LCD).

Example:

$$\frac{2}{3} + \frac{1}{2}$$

You need the LCM of 3 and 2.

3: 3, 6, 9, 12, ...
2: 2, 4, 6, 8, ...
6 is the LCM.

Now you need to make equivalent fractions.

$\frac{2}{3}$ is the same as $\frac{4}{6}$ ($2 \times 2 = 4$ and $3 \times 2 = 6$).

$\frac{1}{2}$ is the same as $\frac{3}{6}$ ($1 \times 3 = 3$ and $2 \times 3 = 6$).

Once the denominators are the same, you just add the numerators.

$\frac{4}{6} + \frac{3}{6} = \frac{7}{6}$. Since $\frac{7}{6}$ is an improper fraction, you may want to rewrite it as a mixed number.

NOTE: To write an improper fraction from a mixed number, you multiply the whole number by the denominator of the fraction. Add this product to the numerator of the fraction. Use this sum as the new numerator over the original denominator.

$\frac{7}{6} = 1\frac{1}{6}$

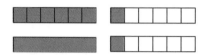

Sometimes you will also need to simplify the answer.

$$\frac{4}{5} + \frac{8}{15}$$

You need the LCM of 5 and 15.

5: 5, 10, 15, . . .

15 is the LCM, and therefore the LCD, of the two fractions.

$\frac{4}{5}$ is the same as $\frac{12}{15}$ ($4 \times 3 = 12$ and $5 \times 3 = 15$).

$\frac{12}{15} + \frac{8}{15} = \frac{20}{15}$. Since $\frac{20}{15}$ is an improper fraction, you may want to rewrite it as a mixed number.

$\frac{20}{15} = 1\frac{5}{15} = 1\frac{1}{3}$

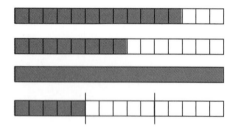

For mixed numbers, you add the fractional parts with the fractional parts and the whole parts with the whole parts, then simplify if needed.

$1\frac{1}{2} + 2\frac{3}{4}$

$1 + 2 = 3$

$\frac{1}{2} + \frac{3}{4} = \frac{2}{4} + \frac{3}{4} = \frac{5}{4}$

Now we have 3 and $\frac{5}{4}$.

Rewrite the $\frac{5}{4}$ as a mixed number: $1\frac{1}{4}$.

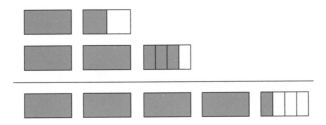

Altogether we have 3 and $1\frac{1}{4}$, so the sum is $4\frac{1}{4}$.

EXAMPLE 2.2   Find the sum:

1) $\frac{5}{14} + \frac{8}{14}$

2) $\frac{7}{12} + \frac{3}{4}$

3) $\frac{4}{9} + \frac{1}{2}$

4) $\frac{3}{8} + \frac{5}{12}$

5) $1\frac{3}{4} + 2\frac{3}{8}$

6) $3\frac{1}{2} + 5\frac{3}{4}$

SOLUTIONS

1) $\frac{13}{14}$

2) $\frac{16}{12}$ or $1\frac{1}{3}$

3) $\frac{17}{18}$

4) $\frac{19}{24}$

5) $4\frac{1}{8}$

6) $9\frac{1}{4}$

2.3 Subtracting Fractions

Remember, adding fractions is a matter of counting; subtraction is adding the opposite or eliminating. If you have $\frac{5}{9}$ and your friend doesn't have any, she may want to borrow $\frac{2}{9}$.

Remember, $\frac{5}{9}$ is just $\frac{1}{9}, \frac{1}{9}, \frac{1}{9}, \frac{1}{9}$, and $\frac{1}{9}$ to make $\frac{5}{9}$. So you have to give your friend $\frac{1}{9} + \frac{1}{9}$; you will have $\frac{3}{9}$ left after she takes the two she wanted to borrow.

This is rather simple when you have the same sized pieces or the same denominator, because you are just counting up your remaining pieces after the subtracted ones are gone.

$$\tfrac{5}{12} - \tfrac{4}{12}$$

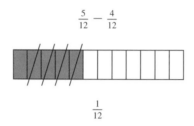

$$\tfrac{1}{12}$$

Similar to addition, when fractions are being subtracted you need to have like denominators. Once the denominators are the same, you just subtract the numerators.

$$\tfrac{3}{4} - \tfrac{5}{12}$$

The LCM is 12.

$\tfrac{3}{4}$ is the same as $\tfrac{9}{12}$ ($3 \times 3 = 9$ and $4 \times 3 = 12$)

$$\tfrac{9}{12} - \tfrac{5}{12}$$

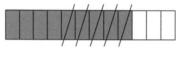

$$\tfrac{4}{12} = \tfrac{1}{3}$$

For mixed numbers, you subtract the fractional parts from the fractional parts and the whole parts from the whole parts, and then simplify if needed. Sometimes, you may need to "borrow" in order to subtract the fractional parts. You "borrow" from the whole parts. Again, this is all about fractions equal to one whole.

$$3\tfrac{1}{2} - 2\tfrac{1}{4}$$
$$3\tfrac{2}{4} - 2\tfrac{1}{4}$$

$$3 - 2 = 1$$
$$\tfrac{2}{4} - \tfrac{1}{4} = \tfrac{1}{4}$$

Put it together for $1\tfrac{1}{4}$

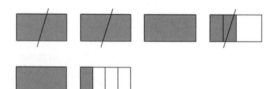

$$5\tfrac{5}{12} - 3\tfrac{3}{4}$$
$$5\tfrac{5}{12} - 3\tfrac{9}{12}$$

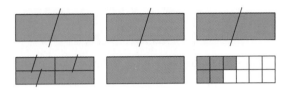

$\frac{5}{12}$ is not enough to subtract $\frac{9}{12}$ from, so we need to regroup from the "wholes."

$\frac{12}{12}$ is the same as 1 whole. So we need to regroup the 5 into 4 and $\frac{12}{12}$, and then add it to the $\frac{5}{12}$.

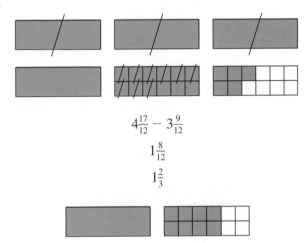

$$4\frac{17}{12} - 3\frac{9}{12}$$

$$1\frac{8}{12}$$

$$1\frac{2}{3}$$

EXAMPLE 2.3

1) $\frac{7}{8} - \frac{3}{8}$

2) $\frac{11}{12} - \frac{1}{2}$

3) $\frac{8}{9} - \frac{2}{3}$

4) $4\frac{7}{8} - 2\frac{3}{4}$

5) $5\frac{1}{2} - 3\frac{3}{4}$

6) $7\frac{1}{2} - 2\frac{3}{5}$

SOLUTIONS

1) $\frac{4}{8}$ or $\frac{1}{2}$

2) $\frac{5}{12}$

3) $\frac{2}{9}$

4) $2\frac{1}{8}$

5) $1\frac{3}{4}$

6) $4\frac{9}{10}$

2.4 Multiplying Fractions

When multiplying fractions, they do NOT have to have like denominators.

$$\frac{2}{3} \times \frac{1}{2}$$

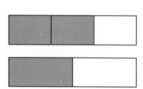

This should be reasonable to you, as $\frac{1}{3}$ and $\frac{1}{3}$ are $\frac{2}{3}$, so $\frac{1}{2}$ of $\frac{2}{3}$ is $\frac{1}{3}$. Remember, this all has to relate back to repeated addition.

Multiply numerator with numerator and denominator with denominator. It is easier if you simplify before you multiply. You can simplify last, but then you are working with larger numbers. Mixed numbers need to be changed to improper fractions.

Examples:

$$\frac{2}{3} \times \frac{9}{10}$$

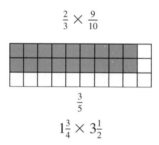

$$\frac{3}{5}$$

$$1\frac{3}{4} \times 3\frac{1}{2}$$

$\frac{7}{4} \times \frac{7}{2}$ Change to improper ($1 \times 4 + 3 = 7$, therefore $\frac{7}{4}$)

$\frac{49}{8}$ Simplify   ($3 \times 2 + 1 = 7$, therefore $\frac{7}{2}$)

$6\frac{1}{8}$

NOTE: To write a mixed number from an improper fraction, you divided the numerator by the denominator. The quotient is the whole number; the remainder is left over the original denominator as the fractional part.

EXAMPLE 2.4

1) $\frac{4}{5} \times \frac{3}{16}$

2) $\frac{5}{6} \times \frac{3}{4}$

3) $\frac{4}{7} \times \frac{3}{8}$

4) $1\frac{1}{2} \times \frac{7}{8}$

5) $1\frac{1}{4} \times 2\frac{3}{5}$

SOLUTIONS

1) $\frac{3}{20}$

2) $\frac{5}{8}$

3) $\frac{3}{14}$

4) $1\frac{5}{16}$

5) $3\frac{1}{4}$

2.5 Dividing Fractions

When dividing fractions, they do NOT have to have like denominators.

$$\frac{2}{3} \div \frac{1}{2}$$

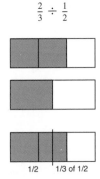

1/2   1/3 of 1/2

This should be reasonable to you, as $\frac{1}{3}$ and $\frac{1}{3}$ are $\frac{2}{3}$—therefore breaking it into two sets of $\frac{2}{3}$ is $\frac{4}{3}$. Remember, this all has to relate back to repeated addition.

Multiply by the reciprocal of the divisor (think back to multiplicative inverses). Then follow the same concepts from multiplying. Multiply numerator with numerator and denominator with denominator. It is easier if you simplify before you multiply. You can simplify last, but then you are working with larger numbers. Mixed numbers need to be changed to improper fractions.

Example:

$\frac{3}{5} \div \frac{2}{3}$

$\frac{3}{5} \times \frac{3}{2}$ rewrite using the reciprocal

$\frac{9}{10}$ multiply

Example:

$1\frac{1}{4} \div 2\frac{1}{2}$

$\frac{5}{4} \div \frac{5}{2}$ rewrite to improper fractions

$\frac{5}{4} \times \frac{2}{5}$ rewrite using the reciprocal

$$\frac{^1\cancel{6}}{_2\cancel{4}} \times \frac{\cancel{2}^1}{\cancel{6}_1} = \frac{1}{2}$$ reduce by cancelling common factors

$$\frac{1}{2}$$

EXAMPLE 2.5

1) $\frac{3}{4} \div \frac{2}{3}$

2) $\frac{3}{8} \div \frac{3}{4}$

3) $\frac{5}{6} \div \frac{3}{4}$

4) $6\frac{2}{3} \div 4\frac{1}{6}$

5) $9 \div 1\frac{1}{2}$

SOLUTIONS

1) $1\frac{1}{8}$

2) $\frac{1}{2}$

3) $1\frac{1}{9}$

4) $1\frac{3}{5}$

5) 6

2.6 Word Problems

The key to solving any word problem, whether there are fractions or not, is reading the question; you need to focus on the needed information. Use key words to help you determine what operations are needed.

Example:

A bicycle path is $3\frac{3}{4}$ miles long. The path is only $\frac{2}{3}$ completed. How much of the path is completed?

What does the question ask? What operation is needed?

$\frac{2}{3}$ of $3\frac{3}{4}$ multiplication

$\frac{2}{3} \times 3\frac{3}{4}$

$\frac{2}{3} \times \frac{15}{4}$

$\frac{15}{6}$ or $2\frac{1}{2}$ miles.

Example:

A recipe calls for $5\frac{1}{2}$ cups of flour. You only have $2\frac{3}{4}$ cups. How much more flour is needed?

What does the question ask? What operation is needed?

$2\frac{3}{4} +$ how much $= 5\frac{1}{2}$	subtraction
$5\frac{1}{2} - 2\frac{3}{4}$	rewrite to like denominators
$5\frac{2}{4} - 2\frac{3}{4}$	regroup
$4\frac{6}{4} - 2\frac{3}{4}$	subtract
$2\frac{3}{4}$	

You need to get $2\frac{3}{4}$ more cups of flour.

1) Lori bought 6 lbs. of grapes. She used $1\frac{1}{2}$ lbs. on Friday and $2\frac{3}{4}$ lbs. on Saturday. How many pounds of grapes does she have left?

2) Augie bought a piece of wood that was $8\frac{1}{2}$ feet long. He used $\frac{2}{3}$ of the wood to make a shelf. How much wood does he have left?

3) Jim rode $5\frac{1}{2}$ miles on his bike on Monday, $4\frac{3}{8}$ miles on Tuesday, and $6\frac{3}{4}$ miles on Wednesday. How many miles did he ride altogether?

4) Kathy used $3\frac{3}{5}$ yards of fabric for a costume. The fabric costs $12.00 per yard. How much did she spend for the fabric?

SOLUTIONS

1) $1\frac{3}{4}$ lbs. left

2) $2\frac{5}{6}$ feet

3) $16\frac{5}{8}$ miles

4) $43.20

Chapter Review

Add

1) $\frac{3}{4} + \frac{5}{8}$

2) $3\frac{3}{4} + 6\frac{2}{3}$

Subtract

3) $\frac{3}{4} - \frac{5}{8}$

4) $2\frac{5}{8} - 1\frac{1}{2}$

5) $4\frac{1}{3} - 2\frac{7}{12}$

Multiply

6) $\frac{5}{12} \times \frac{3}{5}$

7) $2\frac{1}{2} \times 1\frac{3}{5}$

8) $6 \times 2\frac{3}{4}$

Divide

9) $\frac{1}{6} \div \frac{3}{4}$

10) $2\frac{5}{6} \div \frac{2}{3}$

Solve

11) Matty decided he wants to run a 5K race. He is starting by running $1\frac{3}{4}$ km and adding $\frac{1}{2}$ km each day. How many days will it take him to reach his goal?

12) The chili is supposed to cook for $5\frac{1}{2}$ hours. It has already simmered for $2\frac{2}{3}$ hours. How many more hours does it have to cook?

Decimals

WHAT YOU WILL LEARN

- To read, write, and compare decimals

- To add, subtract, multiply, and divide decimals

SECTIONS IN THIS CHAPTER

- How Do We Read, Write, and Compare Decimals?
- How Do We Add and Subtract Decimals?
- How Do We Multiply Decimals?
- How Do We Divide Decimals?

3.1 How Do We Read, Write, and Compare Decimals?

A decimal is any number in our base-ten number system. The places to the right of the decimal point are less than one. The decimal point is used to separate the "ones" place from the "tenths" place. It is also used to separate dollars and cents. As you move right in a decimal number, each place is divided by ten.

Units	Decimal Point	Tenths	Hundredths	Thousandths	Ten-Thousandths	Hundred-Thousandths	Millionths
	·	$\frac{1}{10}$	$\frac{1}{100}$	$\frac{1}{1000}$	$\frac{1}{10,000}$	$\frac{1}{100,000}$	$\frac{1}{1,000,000}$

Decimals are very useful when you need a precise answer. Not all answers will be whole numbers. If you think about money, it is not very often that coins are not needed for a purchase. We use decimals quite often in our lives.

Reading a decimal is rather simple. You need to read the whole number part first. After the whole number the decimal point is read as the word "and." The part after the decimal is read as a whole number, adding the last decimal place after the number part. For example, 56.25 is read as "fifty-six and twenty-five hundredths."

You read the number, add the word "and" at the decimal point, read the number, and add the place-value of the last digit. It is commonly heard as "fifty-six point twenty-five"; however, that is not mathematically correct.

Number	Read
1.348	One **and** three-hundred forty eight thousandths
12.80593	Twelve **and** eighty-thousand, five hundred ninety-three hundred-thousandths
$25.92	Twenty-five dollars **and** ninety-two cents

To write a decimal, you need to use place value for the powers of ten, making sure you place appropriate zeroes as place holders. Also, since each place value is a power of ten you can also write the numbers as a fraction.

Phrase	Decimal	Fraction
Seven tenths	0.7	$\frac{7}{10}$
Forty-nine hundredths	0.49	$\frac{49}{100}$
Nineteen and fifty-six thousandths	19.056	$19\frac{56}{1,000}$

Remember, the number part to the left of the decimal point is the whole number, and the number part to the right of the decimal point is the decimal or fractional part.

PLACE VALUE AND DECIMALS

Millions	Hundred Thousands	Ten Thousands	Thousands	Hundreds	Tens	Ones	And	Tenths	Hundredths	Thousandths	Ten-Thousandths	Hundred-Thousandths	Millionths
				2	1	8	.	6	4				
				3	4	5	.	0	8				
					2	0	.	1	4	9			
						0	.	5	7	0	8		
					2	4	.	1	8				

When comparing decimals you need to remember the importance of place value. Don't fall into the trap of thinking that if a number looks bigger, it has a greater value; that is not true. The best way to compare is to line up the decimal point, which lines up the place values.

Let's compare 24.18 and 20.149.

Line them up: 24.18
20.149

Compare each place, one at a time, from the left to the right.
24.18
20.149

Both have 2 tens, that is the same.
4 > 0; therefore, 24.18 is greater than 20.149.

Let's compare 0.1462 and 0.149.

Line them up: 0.1462
0.149

Compare each, one place at a time from the left to the right.
0.1462
0.149

Both have 1 tenth; that is the same.
Both have 4 hundredths; that is the same.
9 > 6; therefore 0.149 is greater than 0.1462.

EXAMPLE 3.1

1) Write the phrase for each number
   a) 12.09
   b) 327.0084
   c) $123.94

2) Write the decimal number for each phrase
   a) Six hundred twenty-seven and fifty-four thousandths
   b) Three million, four hundred, and six hundred forty-three millionths
   c) Eighty-four dollars and sixty-two cents
   d) Seven tenths

3) Compare using >, <, or =
   a) 34.007   34.0071
   b) 18.056   18.560
   c) 1.008   0.008
   d) .009   0.009
   e) 123.9804   123.978056

SOLUTIONS

1) a) Twelve and nine hundredths
   b) Three hundred twenty-seven and eighty-four ten-thousandths
   c) One hundred twenty-three dollars and ninety-four cents

2) a) 627.054
   b) 3,000,400.000643
   c) $84.62
   d) 0.7

3) a) <
   b) <
   c) >
   d) =
   e) >

3.2 How Do We Add and Subtract Decimals?

Remember a decimal number has a whole number part and a fractional part. When adding or subtracting decimals, we need to line up the decimal points to ensure each digit is in the correct place value position. Doing this lines up whole number parts as well as the fractional parts, and creates like denominators for every digit.

Let's add 12.34 and 27.25

$$\begin{array}{r} 12.34 \\ +\ \ 27.25 \end{array}$$

Add each column (tens with tens, ones with ones, tenths with tenths, hundredths with hundredths)

39.59

Since none of the columns added up to more than 9, we did not have to regroup or carry over to the next column. It is traditional to add from right to left to help with any regrouping or carrying if needed.

Let's add 2.84 and 7.256.

$$\begin{array}{r} 2.84 \\ +\ \ 7.256 \end{array}$$

You may want to place a zero in the thousandth column to have the same number of decimal digits.

$$\begin{array}{r} 2.840 \\ +\ \ 7.256 \end{array}$$

Add each column from right to left:

$$
\begin{array}{r}
1 \\
2.840 \\
+\ \ 7.256 \\
\hline
10.096
\end{array}
$$

Notice, since $8 + 2 = 10$ we had to regroup and carry the one to the ones column. Remember, $\frac{10}{10} = 1$.

Add $0.058 + 2.04 + 0.7$:

$$
\begin{array}{r}
0.058 \\
2.040 \\
+\ \ 0.070 \\
\hline
2.168
\end{array}
$$
Remember to add zeroes as place holders

Notice $5 + 4 + 7 = 16$; therefore the 6 stays in the hundredths place but the 1 needs to go to the tenths place, because 10 hundredths is 1 tenth.

Add $12.67 and $78:

$$
\begin{array}{r}
\$\ \ 12.67 \\
+\ \ \$\ \ 78.00 \\
\hline
\$\ \ 90.67
\end{array}
$$
Adding the zeroes as place holders

Remember to add decimal numbers; it is all about the lineup. Add extra zeroes to the right of the decimal point if needed. Start at the right and add each column in turn. If you need to carry, remember each place value is worth ten of the place value position to the right. Don't forget to bring down the decimal point for the sum.

Subtracting is similar to adding. It is all about the lineup as well. The big difference is you may need to regroup or borrow. This is the reverse of carrying. You will be going to the left. Remember each place value is worth ten of the place value position to the right, therefore we will borrow in groups of ten.

Subtract 8.09 from 9.12.

$$
\begin{array}{r}
9.12 \\
-\ \ 8.09 \\
\hline
\end{array}
$$
change by regrouping
$$
\begin{array}{r}
9.0^{1}2 \\
-\ \ 8.0\ 9 \\
\hline
1.03
\end{array}
$$

Donald went to the hardware store and bought a box of nails. The nails cost $6.32; he gave the cashier a ten dollar bill. How much change did Donald receive?

$$
\begin{array}{r}
\$10.00 \\
-\ \ \$\ 6.32 \\
\hline
\end{array}
$$
change by regrouping
$$
\begin{array}{r}
\$09.9^{1}0 \\
-\ \ \$\ 6.3\ 2 \\
\hline
\$\ 3.68
\end{array}
$$

Remember to subtract decimal numbers—it is all about the lineup. Add extra zeroes to the right of the decimal point if needed. Start at the right and subtract each column in turn. If you need to borrow, remember each place value is worth ten of the place value position to the right. Don't forget to bring down the decimal point for the difference.

EXAMPLE 3.2

1) Add:

12.034 + 24.2

10.4 + 5.72 + 54.3

$10.10 + $8.08 + $50.99

2) Subtract:

34.098 − 5.842

8.743 − 7.4239

$23.98 − $5.87

3) Kathy went to the grocery store and bought eggs for $3.89 and cheese for $2.49 for her omelet. How much did Kathy spend?

If Kathy gave the cashier $20, how much change would Kathy receive?

SOLUTIONS

1) a) 36.234
   b) 70.42
   c) $69.17

2) a) 28.256
   b) 1.3191
   c) $18.11

3) The total was $6.38. Kathy received $13.62 in change.

3.3 How Do We Multiply Decimals?

When multiplying decimal numbers, the placement of the decimal point in the product is very important. You can use estimation to ensure you are placing the decimal point correctly. There is also a pattern that will give you the correct placement of the decimal point.

Multiply   2.3 by 5.7

Estimation $2 \times 6 = 12$ We are looking for an answer around 12.

```
      2.3
  ×   5.7
```
161	find the partial products
1150	find the partial products
1311	You would place the decimal point between the 3 and the 1. The resulting product would be 13.11. This is a number close to our estimate of 12.

Multiply   45.3 by 2.75

Estimation   45 × 3 = 135   We are looking for an answer around 135.

```
      45.3
  ×   2.75
```
2265	find the partial products
31710	find the partial products
90600	find the partial products
124575	You would place the decimal point between the 4 and the 5. The resulting product would be 124.575. This is a number close to our estimate of 135.

You can also use the pattern to help you place the decimal. Since there are a total of two decimal places in our factors of 2.3 and 5.7, there will be two decimal places in our product of 13.11. Since there are a total of three decimal places in our factors of 45.3 and 2.75, there will be three decimal places in our product of 124.575.

Gold costs $1,661.32 per ounce. How much would 6 ounces of gold cost?

Multiply $1,661.32 by 6

```
  1661.32
  ×     6
  9967.92
```

The 6 ounces of gold would cost $9,967.92.

EXAMPLE 3.3

Multiply:
1) 12.8 by 34.7
2) 15 by 3.76
3) 1.93 by 2.08
4) $ 12.89 by 200
5) If silver costs $31.70 per ounce, how much would 5 ounces of silver cost?

SOLUTIONS

1) 444.16
2) 56.40
3) 4.0144
4) $ 2,578.00
5) $ 158.50

3.4 How Do We Divide Decimals?

When the divisor is a whole number and the dividend is a decimal, you would perform the division as you would as if it were two whole numbers, and then move the decimal point straight up in the quotient.

Example:

$$
\begin{array}{r}
3.42 \\
12\overline{)41.04} \\
-36 \\
\hline
5\ 0 \\
-48 \\
\hline
24 \\
-24 \\
\hline
0
\end{array}
$$

When the divisor is a decimal number, you need to make it whole by using powers of ten. You multiply both the divisor and dividend by the power of ten needed to make the divisor whole. This is equivalent to multiplying by one, which will not change the value. Remember $100/100 = 1$. Once the divisor is whole, you divide as you would two whole numbers, and move the decimal point up into the quotient.

$3.5\overline{)51.1}$ multiply both by 10 $35\overline{)511}$

Divide as you would two whole numbers

$$
\begin{array}{r}
14.6 \\
35\overline{)511} \\
-35 \\
\hline
161 \\
-140 \\
\hline
210 \\
-210 \\
\hline
0
\end{array}
$$

Remember, you need the divisor to be a whole number before you divide. You need to use a power of ten. You multiply both the divisor and the dividend by the same power of ten. Divide the dividend by the whole number to find the quotient.

EXAMPLE 3.4

Divide:

1) $1.5\overline{)5.34}$

2) $.6\overline{)76.2}$

3) $6.2\overline{)1227.6}$

SOLUTIONS

1) 3.56

2) 127

3) 198

Chapter Review

1) Michelle spends $2.50 on popcorn and $1.69 on a hot dog. If she buys a soda for $1.75, how much will her total cost be?

2) Matthew gives the clerk $50.00 for a sweatshirt which costs $38.95. How much change will he receive back?

3) Emily buys 6 books for $2.95 each. What is her total?

4) Richard spent $98.80 on 8 shirts. How much was each shirt?

5) How would you write "twenty-six thousand, four hundred and sixty-two hundredths" in standard form?

6) Order from least to greatest:

12.35, 12.035, 12.53, 15.423

Integers

WHAT YOU WILL LEARN

- Understand that positive and negative numbers are used together to describe quantities having opposite directions or values.

- Apply and extend previous understandings of addition and subtraction to add and subtract rational numbers; represent addition and subtraction on a horizontal or vertical number line diagram.

- Apply and extend previous understandings of multiplying and dividing to multiply and divide rational numbers.

SECTIONS IN THIS CHAPTER

- What Are Integers?
- Adding Integers
- Subtracting Integers
- Multiplying Integers
- Dividing Integers

4.1 What Are Integers?

An integer is a whole number (not a fraction) that can be positive, negative, or zero. If you think of all the whole numbers (0, 1, 2, 3, 4, ...) and then add in their opposites, you will have the integers. Opposites have the same numeral part with different signs (positive or negative). Opposites have the same magnitude or absolute value. Therefore, the numbers 1, 0, −2, and 48 are all integers.

The number line below illustrates integers.

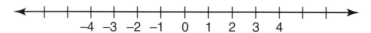

Remember: Integers include negative numbers, zero, and positive numbers. You can omit a positive sign if you choose, but you always need to include a negative sign. Zero is neither positive nor negative.

There are many examples of integers in "real life" situations. Integers can be used for temperature, sea level, gain or loss—even the stock market.

The football team had a five-yard penalty is -5.

You owe your friend a dollar is -1.

You went down seven flights of stairs is -7.

Let's talk about absolute value. Absolute value is the distance from 0 to a number on a number line. This is written symbolically as $|n|$.

opposites

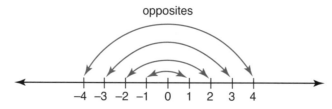

If we look at the number line again, we notice that opposites are the same distance from zero, therefore opposites have the same absolute value. Since absolute value describes a distance, it is never negative.

Knowing how to compare integers is an important skill in the world. We can use the number line to help us start comparing.

smaller ——————————→ bigger

The numbers to the left are always smaller than the numbers to the right.

When comparing two positive numbers, you should be pretty familiar with knowing the bigger number. Just think which number is greater and put the open end of the symbol toward that greater number.

$12 > 3$

A positive number is always greater than a negative number.

$3 > -12$

Think: It is better to have than to owe.

When comparing two negative numbers, you always want less negativity (not a bad idea for the drama in your life as well). Less negativity is better. Better is bigger.

$-5 > -14$

EXAMPLE 4.1

1. Write the sets.
   a) Negative integers
   b) Counting numbers
   c) Non-negative integers

2. Write each expression as an integer.
   a) A gain of 20 lbs.
   b) 25 feet below sea level
   c) Ten degrees below zero

3. Write the opposite.
   a) Going up 6 floors on an elevator.
   b) Losing 19 yards.

4. Compare using $>$, $<$, or $=$.
   a) -12 12
   b) 2 -2
   c) -18 -8
   d) $|-5|$ $|5|$

SOLUTIONS

1. a) $(-1, -2, -3, -4, ...)$

   b) $(1, 2, 3, 4, ...)$

   c) $(0, 1, 2, 3, 4, ...)$ [*remember zero is not negative or positive]

2. a) $+20$

   b) -25

   c) -10

3. a) Going down 6 floors, -6

   b) Gaining 19 yards, $+19$

4. a) -12 $<$ 12
   b) 2 $>$ -2
   c) -18 $<$ -8
   d) $|-5|$ $=$ $|5|$

4.2 Adding Integers

Let's talk about numbers with the same sign first. We would call these "like" signs.

You know how to add two positive numbers. Of course, this isn't going to change.

$3 + 5 = 8$

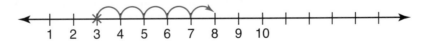

Start at $+3$, move 5 units to the right; you are now at $+8$.

Use 3 pink and 5 pink; altogether, you have 8 pink.

The sum of two positive numbers is always a positive number.

Now let's talk about two negative numbers.

$-3 + -5 = -8$

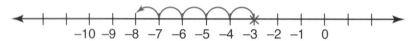

Start at -3, move 5 units to the left; you are now at -8.

Use 3 red and 5 red; altogether, you have 8 red. The sum of two negative numbers is always negative.

You can also relate this to walking down stairs. If you start at the ground level and go down 3 stairs, you are at -3. Now you want to go down 5 more steps; this brings you to -8. You went down a total of 8 stairs. Down results in a negative answer.

In summary, if you are adding two numbers with the same sign, your sum will have the same sign as well. You can use this idea to find the sum using absolute values and then remember to use the sign in your sum.

Now to the integers with unlike signs—we are going to use the number line and integer chips to illustrate again.

$-3 + +5 = +2$

Start at -3, move 5 units to the right; you are now at $+2$.

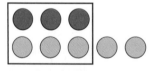

zero pairs

Use 3 red and 5 yellow—you need to pull out your "zero pairs." Opposites make zero pairs. Each red is negative one and zeroes out a yellow, which is positive one. After pulling out the 3 zero pairs, altogether you have 2 yellow.

In both illustrations, 3 negatives and 3 positives resulted in zero. This overlap can be found by subtracting $5 - 3 = 2$. After the overlap was removed, there were 2 positives left over.

In summary, if you are adding two numbers with different signs, your sum will have the sign of the number with the greater absolute value. You can use this idea to find the sum by subtracting the absolute values of the numbers and using the sign of the number with the greater absolute value in your sum.

EXAMPLE 4.2

Add

1) $+3 + -7$

2) $+12 + +13$

3) $-4 + -8$

4) $-17 + -5$

5) $-8 + +4$

SOLUTIONS

1) -4

2) 25

3) -12

4) -22

5) -4

4.3 Subtracting Integers

We are going to again look at the number line and the integer chips. The first (and probably most important) thing is to know that subtraction is adding the opposite. You can relate subtraction to saying "no" to something. If someone says, "Add 5," subtraction would be the opposite: take away 5.

Start with subtraction you already know.

$8 - 5 = 3$ Remember, subtraction is adding the opposite.

$8 - 5 = 8 + (-5)$ change subtracting 5 to adding a negative 5

Start at 8, move 5 units to the left; you are now at $+3$.

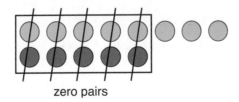

zero pairs

Use 8 pink and 5 red; you need to pull out your "zero pairs." After pulling out the 5 zero pairs, altogether you have 3 pink.

$-3 - (-5) = 2$ Remember, subtraction is adding the opposite.

$-3 - (-5) = -3 + +5$ we change subtracting negative 5 to adding positive 5.

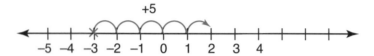

+5

Start at -3, move 5 units to the right; you are now at $+2$.

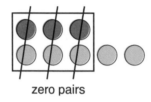

zero pairs

Use 3 red and 5 pink, then pull out your "zero pairs." After pulling out the 3 zero pairs, altogether you have 2 pink.

In summary, if you are subtracting, just change it to an addition problem and follow the rules from section 4.2. Remember, subtraction is adding the opposite.

EXAMPLE 4.3   Subtract
1) $+3 - -7$
2) $+12 - +13$
3) $-4 - -8$
4) $-17 - -5$
5) $-8 - +4$

SOLUTIONS
1)   10
2) -1
3)   4
4) -12
5) -12

4.4 Multiplying and Dividing Integers

Multiplication is repeated addition. This means 4 times 6 is simply adding 6 four times.

$6 + 6 + 6 + 6 = 24$

Since we already know how to add signed numbers, let's use this to understand multiplication.

$4 \times 6 = 24$

$6 + 6 + 6 + 6 = 24$

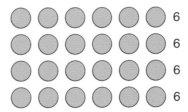

Positive times positive = positive

$4 \times -6 = -24$

$-6 + -6 + -6 + -6 = -24$

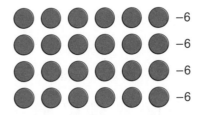

Positive times negative = negative

$-4 \times 6 = -24$

$6 \times -4 = -24$ rewritten using the commutative property

$-4 + -4 + -4 + -4 + -4 + -4 = -24$

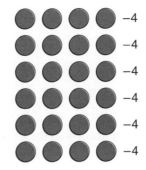

Positive times negative = negative

These two examples are related as $4 \times -6 = -6 \times 4$

In summary, the product of two numbers with different signs will always be negative. Whenever you multiply by a negative number, it changes the sign of the product.

We know the product of two positive numbers will be positive, but what about two negative numbers?

Let's see if this pattern can help.

$-2 \times 3 = -6$
$-2 \times 2 = -4$ increase of 2
$-2 \times 1 = -2$ increase of 2
$-2 \times 0 = 0$ increase of 2   REPEATED ADDITION
$-2 \times -1 = 2$ increase of 2
$-2 \times -2 = 4$ increase of 2

It would appear that negative times a negative is a positive. Let's also look at the integer chips.

$-4 \times -6 = 24$

We can't show "negative sets," so let's start with the 4×-6 we had already done.

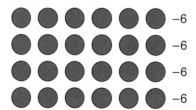

Now I have to use the negative sign, and that would mean flipping the chips to the opposite.

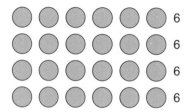

In summary, the product of two numbers with the same signs will always be positive.

Division follows the same rules. Let's take our four examples and "go backwards."

$4 \times 6 = 24$	therefore, $24 \div 6 = 4$
$4 \times -6 = -24$	therefore, $-24 \div -6 = 4$
$-4 \times 6 = -24$	therefore, $-24 \div 6 = -4$
$-4 \times -6 = 24$	therefore, $24 \div -6 = -4$

If you have more than two numbers, you will want to group them into pairs and then multiply or divide.

Example:

$-3 \times -4 \times -5$

$(-3 \times -4) = 12$

$12 \times -5 = -60$

 EXAMPLE 4.4

1) Find the product.
   a) $-3 \times -4 =$
   b) $-5 \times 12 =$
   c) $3 \times -7 =$
   d) $5 \times -6 =$
   e) $-8 \times -4 =$

2) Use the associative property to group and find the product.
   a) $-3 \times -6 \times -4 =$
   b) $-2 \times -5 \times 7 =$
   c) $3 \times -5 \times -7 =$
   d) $5 \times 0 \times -6 =$

3) Find the quotient.
   a) $14 \div -2 =$
   b) $-28 \div 7 =$
   c) $36 \div 6 =$
   d) $-16 \div 8 =$
   e) $-45 \div -5 =$

SOLUTIONS

1) a)   12
   b) −60
   c) −21
   d) −30
   e)   32

2) a) −72
   b)   70
   c) 105
   d)   0

3) a)   −7
   b)   −4
   c)   6
   d)   −2
   e)   9

Chapter Review

1. Use your knowledge of order of operations and signed numbers to solve the problems.
    a. $32 \div -4 \times 3$
    b. $-16 + 24 \div -2$
    c. $5(-7) - 42 \div -6$
    d. $-6(2 + 1)^2 \div -3$

2. Replace each □ with an integer that will make the sentence true.
    a. $-2 \times □ = -6$
    b. $4 + □ = -8$
    c. $12 \div □ = -4$
    d. $7 - □ = 12$
    e. $□ + 5 = 0$

3. On a stormy winter night at 6 p.m., the temperature was 30°F. The temperature dropped six degrees every hour until midnight. What was the temperature at midnight?

Variables and Expressions

WHAT YOU WILL LEARN

- To write, read, and evaluate expressions in which letters stand for numbers.
- To evaluate expressions at specific values of their variables, as well as expressions that arise from formulas used in real-world problems.
- To perform arithmetic operations, including those involving whole number exponents, in the conventional order.

SECTIONS IN THIS CHAPTER

- Using Symbols
- Writing Expressions
- Simplifying Expressions
- Evaluating Expressions

Coefficient A constant that multiplies a variable (for $3x + 4y$, 3 is the coefficient of x and 4 is the coefficient of y).

Constant A quantity that does not change its value in a given expression or equation; a term without a variable (for $s + 4$, 4 is the constant).

Monomial A polynomial with one term; it is a number, a variable, or the product of a number (the coefficient) and one or more variables ($-\frac{1}{4}$, x^2, $4a^2b$, -1.2, $m^2n^3p^4$)

Binomial An algebraic expression consisting of two terms. Example: $5x + 7$

Trinomial A polynomial with exactly three terms. Example: $3x^2 + 7x - 2$

Polynomial A monomial or the sum of two or more monomials whose exponents are positive. Example: $5a^2 + ab - 3b + 7$

Variable A symbol used to represent a number or group of numbers in an expression or an equation.

5.1 Using Symbols

It is often helpful to use a symbol instead of words. In algebra, a letter is often used to stand for a number. The letter used to stand for a number is called a **variable**. You can use any letter you want to use. Commonly, a, b, c, n, x and y are used, or the first letter of what you are representing. If we are writing an expression about dogs, you may want to use the letter d to help you associate. An expression that results in a number is a numeric expression. An expression that contains a variable is an algebraic expression. An algebraic expression can have the variable substituted and become a numeric expression. An expression does not contain any relation symbols such as $<$, $>$, $\leq$, $\geq$, $=$, or $\neq$

$x + 2$	x is the unknown or variable; 2 is the constant
$3y$	3 is the coefficient, y is the variable

Using symbols is like writing in shorthand. Capital letters are reserved for formulas.

$A = lw$	The capital A represents the area. The l is for length and the w is for width. This is much shorter than having to write out "area of a rectangle is the product of the length and the width."
$V = e^3$	The capital V is for volume. The e is for edge. Again, this is much shorter than writing out "volume of a cube is the cube of the edge."

EXAMPLE 5.1 Tell whether the term is a monomial, binomial, or trinomial. Classify as coefficient, variable, exponent, or constant.

1) $2x$
2) $2x^3$
3) $3x + 7$
4) $4x^2 + 21x - 3$
5) $4a^2b^3$
6) $5a - b$

SOLUTION

1) Monomial, 2 is a coefficient, x is a variable
2) Monomial, 2 is a coefficient, x is a variable, 3 is an exponent
3) Binomial, 3 is a coefficient, x is a variable, 7 is a constant
4) Trinomial, 4 and 21 are coefficients, x is a variable, 2 is an exponent, -3 is a constant
5) Monomial, 4 is a coefficient, a and b are variables, 2 and 3 are exponents
6) Binomial, 5 is a coefficient, a and b are variables

5.2 Writing Expressions

Expressions are written using variables, numbers, and operation signs. More complicated expressions will contain parentheses or other grouping symbols. It is important to know key phrases for the operations. Many times words have the same meaning whenever they are used in a statement. However, be careful to always look for context clues and negations. Also, be mindful of the **prepositions** used and the **placement of commas**.

Addition	Subtraction	Multiplication	Division
add	decreased by	multiplied by	divide
added to	difference	multiply	divided by
exceeds by	difference between	product	quotient
increased by	diminished	cubed	half of
more than	less than	squared	halved
plus	lowered by	times	
raised by	minus	twice	
sum	subtract		
total	subtracted from		
	take away		

For multiplication, it is best to use no symbol to avoid confusion with the variable x, or you can use parentheses or a raised dot. For division, it is best to use a fraction. You should always define the variable you are choosing to represent the unknown. For the examples, we will let $n = $ the number.

5 more than a number	$n + 5$
1 less than 6 times a number	$6n - 1$
The quotient of 3 and a number	$\frac{3}{n}$
The difference of 5 and a number	$5 - n$
Three divided by a number increased by 7	$\frac{3}{n + 7}$
Twice the sum of a number and 12	$2(n + 12)$
A number increased by 1, 6 times	$6(n + 1)$
6 times a number increased by 1	$6n + 1$

The placement of the comma indicates order.

EXAMPLE 5.2

Write the expressions using algebra. Let $x =$ the number.
1) twice a number increased by seven
2) a number increased by seven, doubled
3) a number diminished by twelve
4) twelve diminished by a number
5) ten more than a number
6) a number more than ten
7) three times a number decreased by eight
8) eight decreased by three times a number

SOLUTION
1) $2x + 7$
2) $2(x + 7)$
3) $x - 12$
4) $12 - x$
5) $x + 10$
6) $10 + x$
7) $3x - 8$
8) $8 - 3x$

5.3 Simplifying Expressions

When an algebraic expression contains several terms, the terms are separated by $+$ and $-$ signs. These signs belong to the term they precede. In the beginning of an expression, the $+$ is not needed. To "simplify" expressions means to combine the like terms.

Like terms Terms that have the same literal part, same variable raised to the same power.

$2x + 5$ and $3x + 6$ represented using algebra tiles.

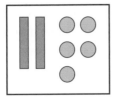 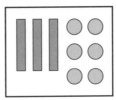 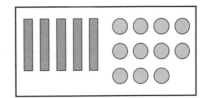

Simply rearrange the tiles so like are with like.

$5x + 11$

$4x - 3$ and $-2x - 5$ represented with algebra tiles.

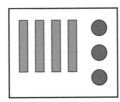  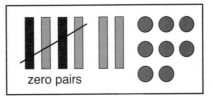

Simply rearrange the tiles so like are with like.

$2x - 8$

In the expression:

$3x^2 + 4x + 5 + 2x + 4x^3 + 5x^2$

$3x^2$ and $5x^2$ are like terms
$4x$ and $2x$ are like terms

Like terms can be combined by adding or subtracting the coefficients and using the commutative, distributive, and associative properties.
The expression would simplify to $4x^3 + 8x^2 + 6x + 5$.

EXAMPLE 5.3

Simplify:
1) $12x + 5 + 4x - 16$
2) $6x + 7 - 3x + 15$
3) $-2x + 6 - 3x + 5$
4) $12ab + -7ab$
5) $8ab^2 + 5ab^2$
6) $4x^2 + 7x - 12 + 2x - 18$

SOLUTIONS
1) $16x - 11$
2) $3x + 22$
3) $-5x + 11$
4) $5ab$
5) $13ab^2$
6) $4x^2 + 9x - 30$

5.4 Evaluating Expressions

Evaluating an expression is all about the substitution property of equality—like if you were to return an item for the exact same priced item in a different color. If the two quantities are equal, then one quantity can replace the other.

Evaluate   To find the value of a mathematical expression.

Substitute   To replace variables in a given expression or equation with designated values in order to evaluate the expression or equation.

Examples:

Find the value of the expressions given $x = -3, y = 4$

$2x^2 - 7$

$2(-3)^2 - 7$ substitute the x with -3—when you plug it, hug it

$2(9) - 7$ follow order of operations

$18 - 7$

11

$3x - 2y$

$3(-3) - 2(4)$ substitute

$-9 - 8$ follow order of operations

-17

$\dfrac{3y}{x}$

$\dfrac{3(4)}{(-3)}$ substitute

$\dfrac{12}{-3}$ follow order of operations

-4

Sometimes you need to simplify before you can evaluate the expression for the given values.

Example:

$5x - 3y + 2x + 8y$ when $x = 4$ and $y = -3$

$7x + 5y$ combine the like terms

$7(4) + 5(-3)$ substitute

$28 + -15$ follow order of operations

13

$-2x + 7x - 4x + 8x$ when $x = 3$

$9x$ combine the like terms

$9(3)$ substitute

27

Evaluate when $a = 4$; $b = 3$; $x = -2$; $y = 5$

1) $2a + 3b$

2) $-5x - 6$

3) $3a^2 + 4b$

4) $\frac{7y}{5x}$

5) $\frac{1}{2}ab^2$

6) $5a - 6b + 3 + 4a + 8b - 12$

7) $7x^2 - 3x + 7 - 4x^2 + 8x$

SOLUTIONS

1) 17

2) -18

3) 60

4) -3.5

5) 18

6) $9a + 2b - 9$; 33

7) $3x^2 + 5x + 7$; 9

Chapter Review

Combine the like terms.

1) $-17xyz - xyz$

2) $24x^2y - 12x^2y$

3) $18ab^2 - 15ab^2$

4) $6a^3b^4 - (-3a^3b^4)$

5) $3x^2 + 7x - 5x^2 + 2x - 8$

6) $19x^2 + 2y - 12y + 6$

7) Why are $11xyz$ and $-35xy$ not like terms?

Translate each phrase into an algebraic expression. Assign the variable d for the unknown. Then, evaluate the above expressions when $d = 8$.

8) Four times a number decreased by 6

9) Eight more than the amount you already have

10) Five runs less than the Yankees scored

11) The quotient of a number and four, minus five

12) Seven increased by the product of the number and -5

Ratios and Proportions

WHAT YOU WILL LEARN

- The concept of a ratio and how to use ratio language to describe a ratio relationship between two quantities.
- How to use ratio and rate reasoning to solve real-world and mathematical problems.
- How to recognize and represent proportional relationships between quantities.
- How to compute unit rates.
- How to use proportional relationships to solve multistep ratio and percent problems.

SECTIONS IN THIS CHAPTER

- What Is a Ratio?
- What Is a Rate?
- Comparing Unit Rates
- What Is a Proportion?
- Solving Proportions
- Similar Figures
- Scale Drawings
- Word Problems

Map scale A key that provides equivalence between a distance on a map and the associated real-world distance.

Means of a proportion The two middle terms in the ratios of a proportion.

Proportion An equation which states that two ratios are equivalent.

Rate A ratio that compares quantities of different units.

Ratio A comparison of two numbers or two like quantities by division.

Scale (1) The ratio of the size of an object in a representation (drawing) of the object to the actual size of the object; the ratio of the distance on a map to the actual distance (e.g., the scale on a map is 1 inch:10 miles); (2) an instrument used to measure an object's mass.

Scale drawing A proportionally correct drawing (enlargement or reduction) of an object or area.

Similar triangles Triangles that have the same shape but not necessarily the same size; corresponding sides are in proportion and corresponding angles are congruent.

Example:

Unit price The price of one item or one unit (e.g., $0.15 per pound).

6.1 What Is a Ratio?

A *ratio is a comparison of two numbers or two like quantities by division*. Ratios can be written in a variety of ways. The main idea is comparison. Think back to a time when your parents were trying to win an argument with you. They might have said, "That is like comparing apples to oranges." Well, in math, we can compare apples to oranges.

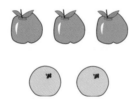

There are three apples and two oranges. The ratio of apples to oranges is 3:2.

When you make a pitcher of ice tea from a powder, there are directions on the container that give you a ratio: You use two scoops of powder per quart. The ratio can be written in three ways:

- Using words (2 to 1)
- Using a colon (2:1)
- Using fractional notation $\left(\frac{2}{1}\right)$

All of these are read, "two to one."

The illustration below shows a ratio of blue paint to yellow paint in a container of green paint. In order to get the same shade of paint, you need to follow the same ratio.

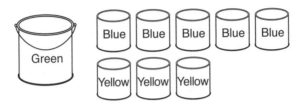

5 blue   3 yellow

5 to 3, 5:3, $\frac{5}{3}$

A ratio is usually written as a fraction in simplest form. For example, the ratio of 30 girls to 20 boys would more simply be stated as $\frac{3}{2}$. A ratio does not have to be a proper fraction and would never be written as a mixed number.

Remember, with ratios, order matters! A ratio of 3:1 is not the same as a ratio of 1:3.

NOTE: To write a ratio in simplest form, you divide both the numerator and denominator by the GCF (Greatest Common Factor, see page 89).

Example:

$\frac{3}{9} = \frac{1}{3}$ Divide both the numerator and denominator by 3.

Example:

Marilyn paid $77.70 for 6 yards of fabric.

$\frac{77.70}{6} = \frac{38.85}{3} = \frac{12.95}{3}$

She paid $12.95 per yard of fabric.

EXAMPLE 6.1

Express each ratio as a fraction in simplest form.
1) 5 rap songs to 25 country songs
2) 12 hits out of 30 times at bat
3) 35 sales out of 70 customers
4) 18 miles in 63 minutes
5) 12 caramels out of 36 chocolates

SOLUTIONS

1) $\frac{1}{5}$ 4) $\frac{2}{7}$

2) $\frac{2}{5}$ 5) $\frac{1}{3}$

3) $\frac{1}{2}$

6.2 What Is a Rate?

A ratio which uses different units of measure is called a rate. You are familiar with a lot of rates; if you have ever been pulled over for speeding, then you are very familiar with miles per hour. Another example, important when purchasing a car, is miles per gallon. You will follow the same concept of simplifying—however, it is very important to keep the unit of measure throughout the computation.

Examples:

A jaguar can run at a speed of 64 miles per hour.

A small hummingbird's wings beat over 100 times per second.

A TV show contains over 20 commercials per hour.

When a rate is simplified so that is has a denominator of 1, it is called a *unit rate*. An example of a unit rate is 15 cents per buffalo wing. Unit rates can be used to comparison shop.

Examples:

$5.99 per pound of turkey breast

$.25 per pencil

$3.29 per gallon of gasoline

There are different ways to find unit rates. One way is to simplify the fraction until the denominator is one. Another way is to divide the cost by the quantity.

Example:

Thomas paid $7.80 for a 65-minute phone call on his cell phone.

$$\frac{\$7.80}{65} = \$.12$$

He paid 12 cents per minute.

Find the unit rate.
1) $4.50 for 12-oz. bottle
2) $29.70 for 6 CDs
3) 238 miles on 7 gallons of gasoline
4) 180 flowers for 5 centerpieces
5) $7.12 for 8 cans of tuna

SOLUTIONS
1) $0.375 per ounce
2) $4.95 per CD
3) 34 miles per gallon
4) 36 flowers per centerpiece
5) $0.89 per can

6.3 Comparing Unit Rates

Sometimes, it is difficult to decide which product is a better value when shopping. This is a good time to use unit price. Most supermarket shelves will have stickers showing the unit price. When comparing unit rate for bargains, the lowest priced item is the better bargain.

Examples:
Which is the better bargain?
7 pens for $1.26 or 9 pens for $1.71

First divide to find the unit price of each item, then compare.
If you purchase the 7 pens, the cost per pen is $.18, but the cost per pen for 9 pens is $.19—therefore, buying the 7 pens is the better bargain.

Which is the better bargain for sunscreen?
8 ounces for $14.24 or 12 ounces for $21.72

First divide to find the unit price of each item, then compare.
If you purchase the 8-ounce bottle, the cost per ounce is $1.78, and the cost per ounce for the 12-ounce bottle is $1.81—therefore, buying the 8-ounce bottle is the better bargain.

Find the better bargain.
1) 3 lbs. of apples for $2.67 or 5 lbs. of apples for $4.30
2) 20 lbs. of dog food for $14.99 or 10 lbs. of dog food for $7.48
3) 4 oz. of perfume for $22.74 or 3 oz. of perfume for $17.58

SOLUTIONS
1) 5 lb. bag
2) 10 lb. bag
3) 4 oz. perfume

6.4 What Is a Proportion?

A proportion is an equation which states that two ratios are equivalent. Remember, a ratio is one quantity compared to another. When we compare two ratios, we do so using proportions.

When you write a proportion, you can use two fractions in an equation or you can use colons.

$$\frac{1}{2} = \frac{2}{4} \qquad \text{or } 1{:}2 = 2{:}4$$

They mean the same thing, but most mathematicians prefer fractions.

Fractions are familiar, so sometimes just by looking at the equation you can tell whether or not it is a proportion. Think back to equivalent fractions when you were younger. Here are some common examples:

$$\frac{1}{2} = \frac{2}{4}$$

$$\frac{1}{3} = \frac{2}{6}$$

$$\frac{1}{4} = \frac{2}{8}$$

$$\frac{1}{10} = \frac{10}{100}$$

Proportional reasoning is very useful. When cooking for a larger group than usual, a chef will use proportional reasoning to double or triple a recipe. Architects use proportional reasoning to create models and scale drawings.

Are these proportions?

The given equation for a proportion must always be true. You can also check if an equation is a proportion by cross-multiplying.

To remember cross-multiplying: think about doing an exercise when you cross your left hand to your right foot and stretch, then cross your right hand to your left foot and stretch, it is the same idea. Your waist is the equal sign. The numerator from the right multiplies with the denominator from the left. The numerator from the left multiplies with the denominator from the right. You may remember the phrase "the product of the means is equal to the product of the extremes" which is similar to cross-multiplying.

Extremes
$$1{:}2 = 2{:}4$$
Means

Examples:

$\frac{1}{2} = \frac{2}{4}$ is true because $1 \times 4 = 2 \times 2$. These are equivalent cross-products.

$3{:}5 = 21{:}35$ is true because $5 \times 21 = 3 \times 35$. These are equivalent cross-products.

Remember, the four parts of the proportion are separated into two groups. You can also check for valid proportions by reducing the fractions to simplest form. Everything is easier to understand when simplified. If the fractions are the same simplified fraction, then they are the same when unsimplified.

$$\frac{4}{16} = \frac{1}{4}$$

$$\frac{7}{28} = \frac{1}{4}$$

Therefore: $\frac{4}{16} = \frac{7}{28}$

EXAMPLE 6.4   Determine if the proportions are valid.

1) $\frac{2}{8} = \frac{3}{9}$

2) $\frac{5}{30} = \frac{2}{12}$

3) $2:5 = 4:25$

4) $5:15 = 7:21$

5) $\frac{1}{9} = \frac{4}{36}$

SOLUTIONS

1) No

2) Yes

3) No

4) Yes

5) Yes

6.5 Solving Proportions

Sometimes, you will be looking for a missing piece of the proportion. Using our knowledge of cross-products being equal, we can solve the equation.

Examples:

$\frac{x}{16} = \frac{3}{12}$

$12x = 48$ cross-products

$x = 4$ divide both sides by 12 to isolate the variable

$\frac{15}{10} = \frac{3}{x}$

$15x = 30$ cross-products

$x = 2$ divide both sides by 15 to isolate the variable

Don't forget you can always "plug in" your answer and check for a valid proportion to verify that your answer is the correct solution.

$\frac{15}{10} = \frac{3}{2}$ because $15 \times 2 = 3 \times 10$ equivalent cross-products

EXAMPLE 6.5   Solve each proportion

1) $\frac{x}{5} = \frac{16}{40}$

2) $\frac{1}{b} = \frac{7}{35}$

3) $\frac{51}{21} = \frac{d}{7}$

4) $\frac{9}{15} = \frac{15}{g}$

5) $\frac{8}{12} = \frac{w}{27}$

6) $\frac{f}{3} = \frac{7.5}{15}$

SOLUTIONS

1) $x = 2$
2) $b = 5$
3) $d = 17$
4) $g = 25$
5) $w = 18$
6) $f = 1.5$

6.6 Similar Figures

Two figures are said to be similar if they have the same shape but not necessarily the same size. Similar triangles have corresponding angles and corresponding sides. When the two figures are similar, the corresponding angles are congruent and the ratios of the lengths of their corresponding sides are equal. You can also say the sides are proportional.

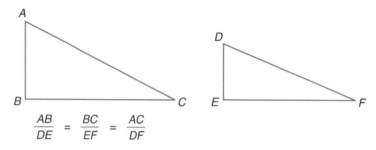

$$\frac{AB}{DE} = \frac{BC}{EF} = \frac{AC}{DF}$$

Therefore, $\triangle ABC \sim \triangle DEF$ (read as $\triangle ABC$ *is similar to* $\triangle DEF$).

The properties of similar triangles are often used for indirect measurement. If you cannot measure the height of an object because it is too tall or too far away, you can use indirect measurement to find the value.

Examples:

If triangle CAT $\sim$ triangle DOG, what is the value of x?

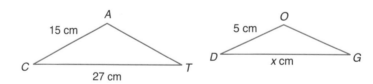

Set up a proportion with corresponding sides

$$\frac{CA}{DO} = \frac{CT}{DG}$$

Plug in the given constants

$$\frac{15}{5} = \frac{27}{x}$$

Use a variable for the unknown
   Solve using cross-products

$$15x = 135$$

Divide by the coefficient

$$x = 9$$

In the figure below, $\triangle ABC \sim \triangle DBF$, find BF.

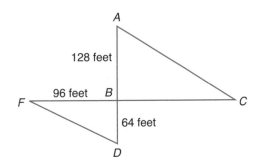

Set up a proportion with corresponding sides

$$\frac{BF}{BC} = \frac{DB}{AB}$$

Plug in the given constants

$$\frac{96}{x} = \frac{64}{128}$$

Use a variable for the unknown
   Solve using cross-products

$$64x = 12{,}288$$

Divide by the coefficient

$$x = 192$$

EXAMPLE
6.6

1) Given $\triangle JKL \sim \triangle MNO$.
   a. List the corresponding angles and corresponding sides.
   b. Find n.

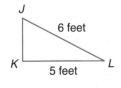
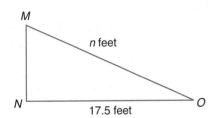

2) Given $\triangle GIH \sim \triangle JIK$, find the missing sides.

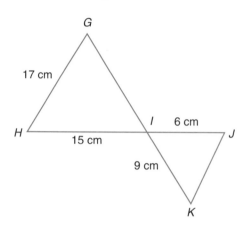

SOLUTIONS

1) a. Angles: J and M, K and N, L and O

   Sides: JK and MN, KL and NO, JL and MO

   b. $n = 21$

2) $JK = 10.2$ and $GI = 10$

6.7 Scale Drawings

A scale drawing is a proportionally correct drawing (enlargement or reduction) of an object or area. Maps, blueprints, and models are examples you should be familiar with from prior experience. It would be impossible and impractical to make a model the actual size of a building. Architects rely on scale models to show their clients the potential of their designs. A photograph is another example of a scale drawing. When you enlarge or reduce a photo, you need to keep the proportions in order to keep the accuracy and not distort the photo. You could also state that the building in the photo and the actual building are similar.

The scale gives the relationship between the measurements on the drawing or model and the measurements on the real object.

Examples

On a map with a scale of 1 inch = 200 miles, what would be the distance between Albany, NY, and Washington, DC, if the actual distance is 360 miles.

Set up a proportion with words
$$\frac{\text{scale}}{\text{actual}} = \frac{\text{scale}}{\text{actual}}$$

Plug in the given constants
$$\frac{1}{200} = \frac{x}{360}$$

Use a variable for the unknown
Solve using cross-products $\qquad\qquad 200x = 360$
Divide by the coefficient $\qquad\qquad\qquad\quad x = 1.8$

The scale distance would be 1.8 inches; however, with inches we would use a fraction, making the answer $1\frac{4}{5}$ inches.

Daniel is planning a garden. His sketch is 3 inches by $4\frac{1}{2}$ inches. If the scale he used is $\frac{1}{2}$ inch = 10 feet, what are the actual dimensions of his garden?

Set up a proportion with words $\qquad\qquad \dfrac{\text{scale}}{\text{actual}} = \dfrac{\text{scale}}{\text{actual}}$

Plug in the given constants $\qquad\qquad\qquad \dfrac{\frac{1}{2}}{10} = \dfrac{3}{x}$

Use a variable for the unknown
Solve using cross-products $\qquad\qquad\qquad \frac{1}{2}x = 30$

Divide by the coefficient $\qquad\qquad\qquad\qquad x = 60$

The width is 60 feet. Now we need to find the length.

Set up a proportion with words $\qquad\qquad \dfrac{\text{scale}}{\text{actual}} = \dfrac{\text{scale}}{\text{actual}}$

Plug in the given constants $\qquad\qquad\qquad \dfrac{\frac{1}{2}}{10} = \dfrac{4\frac{1}{2}}{x}$

Use a variable for the unknown

Solve using cross-products $\qquad\qquad\qquad \frac{1}{2}x = 45$
Divide by the coefficient $\qquad\qquad\qquad\qquad x = 90$

The length is 90 feet. The dimensions of the actual garden are 60 feet by 90 feet. You may choose to use the decimal .5 instead of $\frac{1}{2}$ and 4.5 instead of $4\frac{1}{2}$ when doing your substitution and computation.

EXAMPLE 6.7

1) On a map of California the scale is 1 cm = 50 km. The distance on the map from Los Angeles to San Diego is 3.6 cm. What is the actual distance?

2) The distance from Chicago, IL, to Las Vegas, NV, is 1,500 miles. Using a map scale of 1 inch = 200 miles, what is the scale distance?

3) The scale for a diecast model car is 1:18. The model is 7 inches long. How long is the actual car in feet?

4) A website is selling models of boats ranging in size from $4\frac{1}{2}$ to 18 inches in length. The models are built to a scale of 1:700. What are the smallest and the largest sized actual boats in feet?

SOLUTIONS

1) 180 km

2) $7\frac{1}{2}$ inches

3) $10\frac{1}{2}$ feet

4) $262\frac{1}{2}$ and 1,050 feet

6.8 Word Problems

The hardest part of any word problem is determining which information is necessary, just part of the story, or just there to distract the solver. You should always read the question carefully in order to make a plan. Also, after you have your solution, make sure you go back and check.

Also, be careful with units of measure. Always check to see which unit of measure is being given and which unit of measure is needed for the answer. Many times rates need to be converted, such as minutes to hours or inches to feet. It is a good idea to always write the units throughout the problem.

Examples:

It's a beautiful summer day with a temperature of 80 degrees. Matty wants to go out and enjoy the day, but his friend Steven lives 8 miles away. It takes Matty 45 minutes to ride his bike five miles. At this rate, how long will it take him to get to Steven's house?

Set up a proportion with words
$$\frac{\text{miles}}{\text{minutes}} = \frac{\text{miles}}{\text{minutes}}$$

Plug in the given constants
$$\frac{5}{45} = \frac{8}{x}$$

80 degrees is distracting information.
It is not needed.

Use a variable for the unknown
Solve using cross-products $\qquad$ $5x = 360$
Divide by the coefficient $\qquad$ $x = 72$

It will take Matty 72 minutes to bike to Steven's house.

In Mrs. Paradine's third period class, the ratio of boys to girls is 1:2. If there are 7 boys, how many students are in the class?

First, let's find the number of girls.

Set up a proportion with words $\qquad$ $\dfrac{\text{boys}}{\text{girls}} = \dfrac{\text{boys}}{\text{girls}}$

Plug in the given constants $\qquad$ $\dfrac{1}{2} = \dfrac{7}{x}$

Use a variable for the unknown
Solve using cross-products $\qquad$ $1x = 14$
Divide by the coefficient $\qquad$ $x = 14$
Add the girls and boys together. $\qquad$ $7 + 14 = 21$

There are 21 students in the class.

 EXAMPLE 6.8

1) If the speed of sound is 344 meters per second, what is the approximate speed of sound in meters per minute?
2) Chris ran a distance of 250 meters in two minutes. What is his speed in meters per hour?
3) If a tree 30 feet tall casts a shadow 5 feet long, how tall is a person with a 10-inch shadow?
4) While searching for a college, Julianne found a university with a student to faculty ratio of 1 to 18. If there are 2,500 students, approximately how many faculty members are there?
5) Ella went on a trip to Egypt. The currency conversion is 3.481 Egyptian pounds to one U.S. dollar. How much in U.S. currency would a souvenir for 85 Egyptian pounds cost?

SOLUTIONS

1) 20,640 meters
2) 7,500 meters
3) 5 feet
4) 139 faculty
5) $24.42

Chapter Review

1) James has a bag of marbles. There are 7 red, 6 blue, 3 green, and 5 yellow. Find each of the following ratios in its simplest form.
   a.   Red to blue
   b.   Blue to green
   c.   Green to yellow and red
   d.   Yellow to total marbles

2) Are the ratios 3 to 4 and 6:8 equivalent? Explain your reasoning.

3) Mark runs 5 km in 30 minutes. At this rate, how far could he run in 45 minutes?

4) The instructions for cooking a turkey stated to roast the turkey at 325 degrees for 20 minutes per lb. How many hours will it take to roast a 15-lb. turkey?

5) Sterling silver is made of an alloy of silver and copper at a ratio of 37:3. If the mass of a sterling silver bracelet is 420 grams, how much silver does it contain?

6) Triangle *ABC* has sides with lengths of 3 cm, 5 cm, and 6 cm. Triangle *DEF* has sides of lengths 4 cm, 6 cm, and 8 cm. Determine if the two triangles are similar.

Percents and Percentages

WHAT YOU WILL LEARN

- Understand the concept of a percent and use percent language to describe a relationship between two quantities.
- Find a percent of a quantity as a rate per hundred.
- Solve problems involving finding the whole, given a part and a percent.
- Use percents to solve real-world and mathematical problems.
- Use proportional relationships to solve multistep percent problems.

SECTIONS IN THIS CHAPTER

- What Is a Percent?
- How Do We Calculate a Percent?
- How Do We Find the Whole, Given the Percent?
- How Do We Find the Percent?
- When Is Percent Used?
- What Is Percent Increase or Decrease?
- What Is Percent Error?

7.1 What Is a Percent?

A percent is a special ratio that compares to 100. The root word is cent. There are 100 cents in a dollar.

5% means $\frac{5}{100}$, which is the same as .05

12% means $\frac{12}{100}$, which is the same as .12

We can also reduce the fractions.

$$\frac{5}{100} = \frac{1}{20}$$
$$\frac{12}{100} = \frac{3}{25}$$

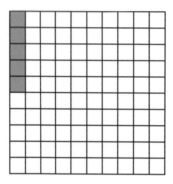 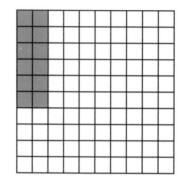

Sometimes, you are given a percent, a decimal, or a fraction. It is useful to know equivalent forms of the same number.

How do you go from a percent to a fraction?

Just remember the 100, write the percent in the numerator and the 100 as the denominator, then reduce.

$$60\% = \frac{60}{100} = \frac{3}{5}$$

How do you go from a percent to a decimal?

Just remember the 100, drop the % sign, and divide by 100

$$60\% = 60 \div 100 = .6$$

How do you go from a decimal to a fraction?

Just remember place value. Read it the RIGHT WAY using correct place value. You always name the place of the last digit. The decimal .6 is read as 6 tenths. The place value of tenths tells you the denominator.

$$\frac{6}{10} = \frac{3}{5}$$

How do you go from a fraction to a decimal?

Just divide the numerator by the denominator.

$$\frac{3}{5} = 3 \div 5 = .6.$$

Of course, common number sense should always prevail. If you know it, do it!!!

EXAMPLE 7.1 · Complete the chart.

Decimal	Fraction	Percent
0.6		
		12.50%
	$\frac{13}{20}$	
0.25		
		18%
	$\frac{1}{2}$	

SOLUTIONS

Decimal	Fraction	Percent
0.6	$\frac{6}{10}$ or $\frac{3}{5}$	60%
.125	$\frac{1}{8}$	12.50%
.65	$\frac{13}{20}$	65%
0.25	$\frac{25}{100} = \frac{1}{4}$	25%
.18	$\frac{18}{100} = \frac{9}{50}$	18%
.5	$\frac{1}{2}$	50%

7.2 How Do We Calculate a Percent?

There are many ways to solve percent problems, but basically it is just multiplying. You can multiply by the decimal or the fraction. You can't multiply by a percent sign. You need to remove the percent sign by changing the percent to an equivalent form. Of course, you might ask why we use different forms of the same number. It all comes down to comfort or convenience. If you owed somebody twenty dollars, what are some ways you can pay them back? You could give them a twenty, two tens, four fives, twenty singles, or any other combination. It is the same idea with percent. You may want to express as a fraction or decimal depending on your comfort level.

IMPORTANT NOTE: Percent can be greater than 100% or even less than 1%.

If you were multiplying by 50, which form of $\frac{2}{25}$ is easiest for you?

$$\frac{2}{25} = \frac{4}{50} = \frac{8}{100} = 8\% = .08$$

If you were multiplying by 200, which form of $\frac{3}{4}$ is easiest for you?

$$\frac{3}{4} = \frac{30}{40} = \frac{75}{100} = 75\% = .75$$

If you were multiplying by 160, which form of $\frac{5}{8}$ is easiest for you?

$$\frac{5}{8} = \frac{10}{16} = \frac{20}{32} = 62.5\% = .625$$

Many people use the patterns to help them calculate as well. These are some quick equivalents.

$$10\% = \frac{1}{10}$$
$$20\% = \frac{1}{5}$$
$$50\% = \frac{1}{2}$$
$$100\% = 1$$
$$200\% = 2$$

SOME COMMON PERCENTAGES

	20	50	130	300
10%	2	5	13	30
20%	4	10	26	60
50%	10	25	65	150
100%	20	50	130	300
200%	40	200	260	600

Do you see the patterns?

10% is half of 20% or 20% is double 10%.
50% is half of 100% or 100% is double 50%.
200% is double 100%, and 100% is always the original total.

Let's try some.

Find 60% of 20.

$$\frac{6}{10} \times 20 = 12 \qquad \text{or} \qquad .6 \times 20 = 12$$

Find 30% of 10.

$$\frac{3}{10} \times 10 = 3 \qquad \text{or} \qquad .3 \times 10 = 3$$

We will change the word problem into either an expression (to be evaluated), an equation, or a proportion (to be solved). Thinking backwards is helpful as well.

For an expression, you want to determine the percent and the original or whole part. You need to write the percent as either a fraction or a decimal. Once you have your decimal or fraction, multiply to get your answer.

Example:

$$30\% \text{ of } 40$$

Expression

$$.3(40) = 12$$
$$\frac{3}{10}(40) = 12$$

For an equation, it is common to use a proportion. This is the same proportional reasoning used to solve similar triangles in Chapter 6. First you want to represent the unknown as a variable and solve. People like to think of it as *part* over *whole* or *is* over *of*.

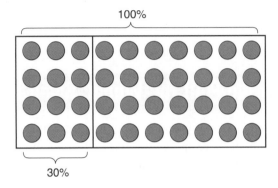

Example:

$$30\% \text{ of } 40$$

Equation/Proportion

$$\frac{\text{is}}{\text{of}} = \frac{\%}{100}$$

$$\frac{x}{40} = \frac{30}{100}$$

$$x = 12$$

More Examples:

What is 90% of 63 ?

Expression with decimal

$$.9(63) = 56.7$$

Expression with fraction

$$\frac{9}{10}(63) = 56.7$$

Proportion

$$\frac{x}{63} = \frac{90}{100}$$

$$x = 56.7$$

Jenna saved 25% of her paycheck. If she gets paid $120, how much does she save?

Rephrase: 25% of 120

Expression with decimal

$$.25\,(120) = 30$$

Expression with fraction

$$\frac{1}{4}(120) = 30$$

Proportion

$$\frac{x}{120} = \frac{25}{100}$$

$$x = 30$$

Rephrase as "she saved $30."

EXAMPLE 7.2

1) Find 75% of 40.
2) Find 85% of 500.
3) Find 5% of 200.
4) Find 16% of 350.
5) Find 25% of 24.

SOLUTIONS

1) 30
2) 425
3) 10
4) 56
5) 6

7.3 How Do We Find the Whole, Given the Percent?

We are going to use equations or proportions to find the whole. It is all about making the connection between the percent and the 100 to the part and the whole. You may need to take your time. Identifying the whole is crucial. Remember:

$$\frac{\text{part}}{\text{whole}} = \frac{\text{percent}}{100}$$

30 is 20% of what number?

The 30 is the part. The whole is unknown.

$$\frac{30}{x} = \frac{20}{100}$$
$$x = 150$$

15 is 75% of what number?

The 15 is the part. The whole is unknown.

$$\frac{15}{x} = \frac{75}{100}$$
$$x = 20$$

168 is 140% of what number?

The 168 is the part. The whole is unknown. Note: the percent is bigger than 100, therefore, the part is bigger than the whole.

$$\frac{168}{x} = \frac{140}{100}$$
$$x = 120$$

Lauren paid $28.80 for a shirt that was on sale for 20% off the original price. What was the original price?

Think: what percent is the $28.80? She saved 20% off of the 100% percent, so $100 - 20 = 80$. She paid for 80% of the dress, so $28.80 is 80% of the original.

$$\frac{28.80}{x} = \frac{80}{100}$$
$$x = 36$$

The original price is $36.

EXAMPLE 7.3

1) 30 is 60% of what number?
2) 52 is 20% of what number?
3) 6 is 5% of what number?
4) 18 is 40% of what number?
5) 75 is 150% of what number?

SOLUTIONS

1) 50
2) 260
3) 120
4) 45
5) 50

7.4 How Do We Find the Percent?

We are going to use equations or proportions to find the percent when given the part and the whole. It is all about making the connection between the percent and the 100 to the part and the whole. You may need to take your time. Identifying the whole is crucial.

Remember:

$$\frac{\text{part}}{\text{whole}} = \frac{\text{percent}}{100}$$

IMPORTANT NOTE: A part can be bigger than the whole if the percent is over 100%.

Three is what percent of 200?

3 is the part, 200 is the whole, and the percent is unknown.

$$\frac{3}{200} = \frac{x}{100}$$
$$x = 1.5$$

Ana is having a party. She has four responses saying they cannot come to the party, and 26 responses saying they will attend. What percent of the invited guests will be attending?

26 is the part, 30 is the whole (26 + 4 = 30), and the percent is unknown.

$$\frac{26}{30} = \frac{x}{100}$$
$$30x = 2600$$
$$x = 86\frac{2}{3}\%$$

Jen bought a shirt that was originally $85.00 but is on sale for $59.50. What percent was saved?

The 85 is the whole. $85.00 − $59.50 = $25.50, which represents the part saved.

$$\frac{25.50}{85} = \frac{x}{100}$$
$$85x = 2500$$
$$x = 30$$

She saved 30%.

1) Thirty-three is what percent of 220?
2) Seventeen is what percent of 85?
3) Lisa saved $64.80 off a jacket which was originally $162.00. What percent did she save?
4) Bart lost 32 lbs. on a new exercise plan. He was originally 256 lbs. What percent of his original weight did he lose?

SOLUTIONS

1) 15%
2) 20%
3) 40%
4) 12.5%

7.5 When Is Percent Used?

Percent is used throughout our lives. Percent is in shopping, banking, budgeting, payroll, and much more.

Here are some examples.

SALES TAX

Sales Tax is a percent charged by the government on a purchase; it differs from location to location.

Sales tax is paid on the amount of your purchase price.
You bought a shirt for $40 and paid 8% sales tax.

Find 8% of 40.

$$.08 \times 40 = 3.2$$

Sales tax is added to your purchase.

$$40 + 3.2 = 43.20$$

You paid a total of $43.20 for your purchase.

Another way is to add the percent of tax to 100% to represent the total percent of the purchase as 108%. You would be adding the percent in the beginning steps of solving the problem.

$$108\%(40) = 43.20$$

You bought a computer for $525 and paid 6% sales tax.

Find 6% of 525.

$$.06 \times 525 = 31.5$$

Sales tax is added to your purchase.

$$525 + 31.5 = 556.5$$

You paid a total of $556.50 for your purchase.

Another way is to add the percent of tax to 100% to represent the total purchase as 106%. You would be adding the percent in the beginning steps of the problem.

$$106\%(525) = 565.5$$

INCOME

The amount of money received for labor, for services, from the sale of goods or property, or from investments.

WITHHOLDING TAX

The amount of payroll tax the government is sent from your employer on your behalf; differs from person to person.

Withholding tax is subtracted from your paycheck. The state and federal government each have a different percent. Some cities also withhold taxes.

You made $500 gross pay and paid 15% federal tax and 10% state tax.

Find 25% of 500 (10% + 15% = 25%).

$$.25 \times 500 = 125$$

Withholding tax is subtracted from your paycheck.

$$500 - 125 = 375$$

You will get a paycheck of $375.00.

Another way is to subtract the percent of tax from 100% to represent the total percent of net pay as 75%. You would be subtracting the percent in the beginning steps of the problem.

$$75\%(500) = 375$$

You made $1,200 gross pay and paid 23% federal tax and 13% state tax.

Find 36% of 1200 (23% + 13% = 36%).

$$.36 \times 1200 = 432$$

Withholding tax is subtracted from your paycheck.

$$1200 - 432 = 768$$

You will get a paycheck of $768.00.

Another way is to subtract the percent from 100% to represent the total percent of net pay as 64%. You would be subtracting the percent in the beginning steps of the problem.

$$64\%(1200) = 768$$

COMMISSION

The amount of earnings based on a percent of sales.

Commission is paid based upon an agreement between the employee and employer. Some employees receive salary in addition to commission and some receive only commission.

You sold $20,000 worth of computer parts and make 5% of your sales.

Find 5% of 20,000.

$$.05 \times 20,000 = 1,000$$

You will receive $1,000 in commission.

You sold $540 worth of cosmetics and make 12% of your sales.

Find 12% of 540.

$$.12 \times 540 = 64.80$$

You will receive $64.80 in commission.

INTEREST

The amount of money charged for borrowing money or the profit (usually money) that is made on invested capital.

INTEREST RATE

The percent of interest charged on money borrowed or earned on money invested.

Interest is paid to you on savings and investments. You pay interest on loans or credit cards. We are going to use a simple interest formula.

$$Interest = (principal)(percent)(time\ in\ years)$$

You put $400 in a CD for 3 years at 4% interest.

Substitute into the formula.

$$I = (400)(.04)(3)$$
$$I = 48$$

You will make $48 in interest.

You put $2,000 in a savings account for 18 years at 2.5% interest.

Substitute into the formula.

$$I = (2000)(.025)(18)$$
$$I = 900$$

You will make $900 in interest.

DISCOUNT

An amount deducted from the gross amount or price of an item or service.

SALE PRICE

The price of a product after the discount has been subtracted from the original price.

Discounts vary as they are set by the store. Usually a percentage, but may be a dollar amount.

Sally's Super Store is having a 20% off sale and you bought a pair of jeans that were originally $80.

Find 20% of 80.

$$.2 \times 80 = 16$$

Discounts are subtracted from your purchase.

$$80 - 16 = 64$$

You paid a total of $64 for your purchase.

Another way is to subtract the percent from 100% to represent the total purchase price as 80%. You would be subtracting the percent in the beginning steps of the problem.

$$80\%(80) = 64$$

Dave's Discounts and Deals is having a 35% off sale and you bought an item that was originally $210.

Find 35% of 210.

$$.35 \times 210 = 73.5$$

Discounts are subtracted from your purchase.

$$210 - 73.5 = 136.5$$

You paid a total of $136.50 for your purchase.

Another way is to subtract the percent from 100% to represent the total purchase price as 65%. You would be subtracting the percent in the beginning steps of the problem.

$$65\%(210) = 136.5$$

GRATUITY

A tip given for good service.

Gratuities are set by the purchaser. It is suggested to give between 15–20% for good wait service.

You spent $25 on dinner and you give a 20% tip.

Find 20% of 25.

$$.20 \times 25 = 5$$

Gratuities are added to your purchase.

$$25 + 5 = 30$$

You paid a total of $30 for your purchase.

Another way is to add the percent to 100% to represent the total purchase price as 120%. You would be adding the percent in the beginning steps of the problem.

$$120\%(25) = 30$$

You spent $180 on dinner and you give an 18% tip.

Find 18% of 180.

$$.18 \times 180 = 32.4$$

Gratuities are added to your purchase.

$$180 + 32.4 = 212.4$$

You paid a total of $212.40 for your purchase.

Another way is to add the percent to 100% to represent the total purchase price as 118%. You would be adding the percent in the beginning steps of the problem.

$$118\%(180) = 212.4$$

 EXAMPLE 7.5

1) You bought a shirt for $80 and paid 7% sales tax. What is your total?
2) You spent $70 on dinner and you give a 20% tip. What is the amount of the tip given?
3) You put $600 in a savings account for 12 years at 3.5% interest. How much interest will you earn?
4) Katie's Kupcakes is having a 25% off sale, and you bought items that were originally $21. How much did you spend?
5) You sold $2,000 of office furniture and make 14% of your sales. What is your commission amount?
6) You made $850 gross pay and paid 18% federal tax and 9% state tax. What is your net pay?

SOLUTIONS
1) $85.60
2) $14
3) $252
4) $15.75
5) $280
6) $620.50

7.6 What Is Percent Increase or Decrease?

Percent decrease   The magnitude of decrease expressed as a percent of the original quantity.

Percent increase   The magnitude of increase as a percent of the original quantity.

Let's start simple: An increase is when something goes up and a decrease is when something goes down. Percent increase or decrease relates the amount of change back to the original amount using percent.

Did the following prices increase or decrease?

1) Original: $40.00
    New: $50.00

2) Original: $50.00
    New: $35.00

3) Original: $800.00
    New: $700.00

If the new price is greater, it is an increase.
    If the new price is less, it is a decrease.
    We will not be using negative signs for a decrease, we use the word decrease instead.

Now, let's go back. How much did they increase or decrease by?

1) increase of $10

2) decrease of $15

3) decrease of $100

The key to calculating the percent increase or decrease is to refer back to the ORIGINAL. It is similar to percent error when you refer back to the ACTUAL, which we will talk about later in this section.
    The percent increase/decrease is the ratio of change compared to the original written as a percent. Finding the amount of change was the first step. Next, write the ratio. Lastly, convert to a percent.

$$\frac{\text{change}}{\text{original}} \times 100$$

Let's look back at our three questions.

1) $\frac{10}{40} \times 100 = 25\%$

2) $\frac{15}{50} \times 100 = 30\%$

3) $\frac{100}{800} \times 100 = 12.5\%$

The $100 decrease is actually the lowest percent of change even though it is the most money. Percent helps us from jumping to conclusions. If the price of an item increases by a few pennies, it still may be a large percentage increase.

You can use this formula.

$$\text{Percent Increase or Percent Decrease} = \frac{\text{New} - \text{Original}}{\text{Original}} \times 100$$

You can also use proportional reasoning.

$$\frac{\text{Difference}}{\text{Original}} = \frac{\text{Percent}}{100}$$

EXAMPLE 7.6

Find the percent increase or decrease.
1) Original: $50.00
   New: $70.00
2) Original: $50.00
   New: $45.00
3) Original: $.30
   New: $.27
4) Original: $25.00
   New: $28.75
5) Original: $250.00
   New: $100.00

SOLUTIONS
1) 40% increase
2) 10% decrease
3) 10% decrease
4) 15% increase
5) 60% decrease

7.7 What Is Percent Error?

When in science you need to compare results of experiments, you will look at the difference in measurements. The absolute difference is not always very helpful, because it depends on the magnitude of the error. If you were measuring the growth of a plant for an experiment, being off by a centimeter is significant; however, if you were measuring the distance from New York to California, it would be of little value.

Relative error   The ratio of the absolute error in a measurement to the size of the measurement; often written as a percent and called the percent of error; the absolute error is the difference between an approximate number and the true number that it approximates.

For percent error, you are comparing your result to the actual result as a ratio expressed as a percent. As with percent decrease or increase, we are only looking for how much we are off by, so we are going to use the absolute value and have no negative answers.

Similar to percent increase or decrease, you find the difference between the actual and the experimental (your result) and express it as a fraction, then multiply by 100 to convert to a percent.

You can use this formula:

$$\% \text{ error} = \frac{\text{your result} - \text{accepted value}}{\text{accepted value}} \times 100$$

A student measures the volume of a 2.50-liter container to be 2.42 liters. What is the percent error in the student's measurement?

$$\% \text{ error} = \frac{2.42 - 2.50}{2.50} \times 100 = 32\%$$

A student takes an object with an accepted mass of 200.00 grams and masses it on his own balance. He records the mass of the object as 196.5 g. What is his percent error?

$$\% \text{ error} = \frac{196.5 - 200}{200} \times 100 = 1.75\%$$

1. A student finds the density of a piece of aluminum to be 2.45 g/cm³. The actual value is 2.699 g/cm³. What is the student's percent error to the nearest tenth?

2. A student made a mistake when measuring the volume of a barrel. He found the volume to be 65 liters. However, the real value for the volume is 50 liters. What is the percent error?

SOLUTIONS

1) 9.2%

2) 30%

Chapter Review

1) Find the percent increase of a stock that was originally $250 and is now $400.

2) Find the percent error to the nearest tenth when the actual weight is 3.65 grams and you measured 3.5 grams.

3) You bought a shirt for $60 and paid 8% sales tax. What is the final cost?

4) You made $1,500 gross pay and paid 15% federal tax and 10% state tax. What is your net pay?

5) You sold $120,000 worth of computer parts and make a commission of 7% on your sales. How much commission would you make?

Factors and Exponents

WHAT YOU WILL LEARN

- How to find all factor pairs for a whole number in the range 1–100; recognize that a whole number is a multiple of each of its factors.
- How to find common factors and multiples.
- How to find the prime factorization of a number.
- How to develop and apply the laws of exponents for multiplication and division.
- How to use scientific notation.
- How to evaluate an expression with integral exponents.

SECTIONS IN THIS CHAPTER

- What Are Factors, Factor Pairs, and Exponents?
- What Is Prime Factorization?
- What Are Common Factors and Multiples?
- What Are the Laws of Exponents?
- What Is Scientific Notation?
- How Can We Evaluate Expressions with Exponents?

8.1 What Are Factors, Factor Pairs, and Exponents?

Factor (noun)   A number or expression that is multiplied by another to yield a product.

Factor (verb)   To express as a product of two or more factors.

When a number is divisible by another number, the smaller number is known as a factor. Factor times factor equals a product. When we are listing factors of a number, we are using counting numbers.

For example, let's try to find the factors of 12. We can find these by dividing. If you divide 12 by 12 the quotient is 1, therefore both 12 and 1 are factors. If you divide 12 by 6, the quotient is 2, therefore both 6 and 2 are factors. If you divide 12 by 4, the quotient is 3; therefore both 4 and 3 are factors. Conversely, if you divide 12 by 5, the quotient is 2.4—12 is not evenly divisible by 5, and therefore 5 is not a factor. The factors of 12 are 1, 2, 3, 4, 6, and 12.

One and the number you are trying to factor are always factors of the number. All the other factors will be greater than one and less than the number. It always seems easier when you pull out the factor with its partner or as a pair. A factor pair is the two numbers which are multiplied together to get the product.

For example, let's look at 18.
We know 1 and 18 are factors.
Moving to 2: 18 divided by 2 is 9; therefore, 2 and 9 make up a factor pair.
Moving to 3: 18 divided by 3 is 6; therefore, 3 and 6 make up a factor pair.
Moving to 4: 18 is not divisible by 4.
Moving to 5: 18 is not divisible by 5.
Moving to 6: 6 is already on the list; you can stop. You have found all the factors

$$18: \quad 1, 2, 3, 6, 9, 18$$

As another example, let's look at 50. We know that 1 and 50 are factors. This will start our list, and we will keep going till we hit a repeat.

$$50: \quad 1, 50 \quad 2, 25 \quad 5, 10$$
Didn't work: 3, 4, 6, 7, 8, 9,
stopped at 10 (already on list)

Looking at exponents

Base   A number that is raised to an exponent.

Example:

2^3, where 2 is the base and 3 is the exponent.

Exponent   A number that tells how many times the base is used as a factor; in an expression of the form b^a, a is called the exponent, b is the base, and b^a is a power of b.

Exponents can be a little tricky. An exponent is like shorthand or code. You need to learn the code to decipher the meaning. You can break exponents down into four groups (we are excluding zero as a base because zero has its own rules).

If the exponent is a counting number, the exponent will tell you how many times to use the base as a factor.

Example:

$$5^3 \text{ means } 5 \times 5 \times 5 \text{ which equals } 125.$$

You can find the value of an exponential expression by multiplying.

$$3^2 = 3 \times 3 = 9$$
$$3^3 = 3 \times 3 \times 3 = 27$$
$$3^4 = 3 \times 3 \times 3 \times 3 = 81$$
$$3^5 = 3 \times 3 \times 3 \times 3 \times 3 = 243$$

When you have a negative number being raised to a power, be sure to include the sign.

$$(-2)^2 = -2 \times -2 = 4$$
$$(-2)^3 = -2 \times -2 \times -2 = -8$$
$$(-2)^4 = -2 \times -2 \times -2 \times -2 = 16$$

When the negative is in front of the expression, only the product is using the negative sign, not the factors; you need parentheses to exclude the negative as part of the factor.

$$-2^3 = -(2 \times 2 \times 2) = -8$$

If the exponent is zero, the base is not being used as a factor. The answer is 1.

Example:

$$5^0 \text{ which equals } 1.$$

If the exponent is negative, the negative sign indicates to use the multiplicative inverse or reciprocal. The number part of the exponent tells you how many times to use the base as a factor.

Example:

$$5^{-3} \text{ means } \tfrac{1}{5} \times \tfrac{1}{5} \times \tfrac{1}{5} \text{ which equals } \tfrac{1}{125}.$$

If the exponent is a fraction, the numerator will tell you how many times the base is used as a factor. The denominator will tell you the root you are trying to find.

Example:

$25^{\frac{1}{2}}$ means square root of 25 which equals 5 because $5 \times 5 = 25$. The square root is the inverse of squaring. You are "going backwards" by asking yourself which number was used to get the product of 25 when multiplied by itself twice.

Example:

$27^{\frac{2}{3}}$ means the square of the cube root of 27. First you find the cube root of 27, which means a number was multiplied by itself three times to get the product of 27. The answer is 3, because $3 \times 3 \times 3 = 27$. Next you would have to square the 3 to follow the 2 of the numerator. 3^2 is 9.

1) List all the factors of:
   a) 24
   b) 72
   c) 25

2) Find the value of the expressions
   a) 2^5
   b) 5^4
   c) 10^2
   d) 6^0
   e) 6^{-2}
   f) 4^{-3}
   g) $81^{\frac{1}{2}}$
   h) $16^{\frac{3}{4}}$
   i) $(-4)^3$
   j) -6^2
   k) -2^4
   l) $(-1)^6$

SOLUTIONS

1) a) 24: 1, 2, 3, 4, 6, 8, 12, 24
   b) 72: 1, 2, 3, 4, 6, 8, 9, 12, 18, 24, 36, 72
   c) 25: 1, 5, 25

2) a) 32
   b) 625
   c) 100
   d) 1
   e) $\frac{1}{36}$

f) $\frac{1}{64}$

g) 9

h) 8

i) −64

j) −36

k) −16

l) 1

8.2 What Is Prime Factorization?

Prime number   A number greater than 1 that has exactly two different factors, 1 and itself.

Examples:

Prime Numbers		Non-Prime Numbers	
Number	Factors	Number	Factors
2	1, 2	6	1, 2, 3, 6
7	1, 7	8	1, 2, 4, 8
11	1, 11	15	1, 3, 5, 15
17	1, 17	25	1, 5, 25

Prime factorization   A method of writing a composite number as a product of its prime factors.

Every counting number greater than 1 is either a prime or a composite number. A prime number has exactly two factors—1 and the number itself. A composite number has at least three factors. Composite numbers can be broken down into their prime parts or prime factors. Every composite number has a unique set of prime factors. We are going to decompose a composite number into its prime factors.

Let's start with 48.

Pick any two factors of 48 other than 1 and 48.

Keep making branches by doing the same process for the factors you chose.

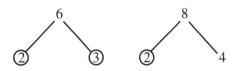

Circle the prime numbers as those are complete.

Once all the branches have stopped with circles you are finished.

Gather your numbers for the solution.

$$48 = 2 \times 2 \times 2 \times 2 \times 3 \quad \text{or} \quad 2^4 \times 3.$$

Another example:

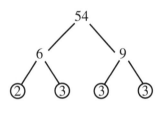

$$54 = 2 \times 3^3$$

You can always check your answer by multiplying.

$$2 \times 3 \times 3 \times 3 = 54$$

 1) Find the prime factorization of the following:
   a) 36
   b) 42
   c) 96

SOLUTIONS
   1) a) $2^2 \times 3^2$
       b) $2 \times 3 \times 7$
       c) $2^5 \times 3$

8.3 What Are Common Factors and Multiples?

Common factor   A number, polynomial, or quantity that evenly divides into two or more numbers or algebraic expressions.

Greatest common factor (GCF)   The greatest number or expression that is a factor of two or more numbers or expressions.

Multiple   The product of a given whole number and any other whole number.

Common multiple   A whole number that is a multiple of two or more given numbers.

Least common multiple (LCM)   The smallest number, greater than zero, which is a multiple of two or more numbers.

Common factors are really just like they sound: factors which are the same. This is often used in reducing fractions. To find a common factor you can look at both lists of factors.

For example:

$$12:\quad 1, 2, 3, 4, 6, 12$$
$$20:\quad 1, 2, 4, 5, 10, 20$$

The common factors are the factors on both lists: 1, 2, 4. The greatest common factor would be 4, since it is the greatest number on both lists.

Another way to find the greatest common factor would be to decompose the numbers to their prime factors.

$$12:\quad 2 \times 2 \times 3$$
$$20:\quad 2 \times 2 \times 5$$

The factors which are common to both lists are 2×2 which you multiply to get the GCF of 4.

Let's try it with three numbers:

$$24:\quad 1, 2, 3, 4, 6, 8, 12, 24$$
$$18:\quad 1, 2, 3, 6, 9, 18$$
$$42:\quad 1, 2, 3, 6, 7, 14, 21, 42$$

The common factors are 1, 2, 3, and 6. The GCF is 6 as it is the greatest number.

Let's try it using prime factors.

$$24:\quad 2 \times 2 \times 2 \times 3$$
$$18:\quad 2 \times 3 \times 3$$
$$42:\quad 2 \times 3 \times 7$$

The numbers that are the same on all the lists are 2 and 3; therefore, the GCF is 6 because $2 \times 3 = 6$.

Finding a multiple is easier than finding a factor, because all you need to do is multiply. Start with the given number and then multiply by 1, 2, 3, 4, 5, etc. Multiples will always be greater than or equal to the given number. The list of multiples is neverending.

Multiples of 4:   4, 8, 12, 16, 20, . . .
Multiples of 7:   7, 14, 21, 28, 35, 42, . . .

When you are looking for a common multiple, you need to compare your lists. You may have to write down a lot of multiples until you find a common one so we are going to start with the larger number.

Multiples of 7:   7, 14, 21, 28, 35, 42, 49, 56, 63, 70, 77, 84, 91, 98, 105, 112, . . .
Multiples of 4:   4, 8, 12, 16, 20, 24, 28, 32, 36, 40, 44, 48, 52, 56, 60, 64, 68, 72, 76, 80, 84, 88, 92, 96, 100, 104, 108, 112, . . .
Multiples of both 4 and 7:   28, 56, 84, 112

28 is the least common multiple because it is the lowest. You would normally stop listing the multiples once you find a match.

This can also be done using prime factorization.

Prime factorization of 4: 2×2
Prime factorization of 7:   7

Since there are no common factors, you would multiply $2 \times 2 \times 7 = 28$.

Let's look at one where there are common factors.

12:   $2 \times 2 \times 3$
15:   3×5

Use the factors $2 \times 2 \times 5$ and then only one of the 3's because the three is repeated. You would use the most repeated occurrence. You would compute $2 \times 2 \times 3 \times 5$ for the LCM of 60.

Let's look at one using three numbers.

Multiples of 8:   8, 16, 24, 32, 40, 48, 56, 64, 72
Multiples of 9:   9, 18, 27, 36, 45, 54, 63, 72
Multiples of 12:   12, 24, 36, 48, 60, 72

72 is the LCM.
Using prime factorization:

8:   $2 \times 2 \times 2$
9:   3×3
12:   $2 \times 2 \times 3$

Use $2 \times 2 \times 2 \times 3 \times 3$, since they are the most repeated occurrences; the LCM is 72.

1) Find the GCF:
   a) 36 and 24
   b) 18 and 100
   c) 121 and 44
   d) 12, 28, and 32

2) Find the LCM.
   a) 9 and 12
   b) 4 and 10
   c) 18 and 22
   d) 12, 15, and 18

SOLUTIONS

1) a) 12
   b) 2
   c) 11
   d) 4

2) a) 36
   b) 20
   c) 198
   d) 180

8.4 What Are the Laws of Exponents?

Law of exponents for multiplication   The product of two or more numbers in exponential form with the same base is equal to that base raised to the power equal to the sum of the powers of each number.

To multiply numbers with exponents that have the SAME base, add the exponents.

Examples:

$4^3 \times 4^5$

Means $4 \times 4 \times 4 \times 4 \times 4 \times 4 \times 4 \times 4$

Which is the same as 4^8

You can get the same answer by adding the exponents of 3 and 5 to get the exponent of 8.

$8^3 \times 8^{-2}$

Means $\dfrac{8 \times 8 \times 8}{8 \times 8}$

Which is the same as 8 or 8^1

You can get the same answer by adding the exponents of 3 and -2 to get the exponent of 1.

To multiply numbers with exponents that have a DIFFERENT base, try to make them the same base, if possible.

If you cannot make them into numbers with the same base, then you are just writing them together as one term.

Example:

$3^4 \times 4^2$

Would stay as $3^4 \cdot 4^2$

If you can make them the same, then first make the bases the same, and then add the exponents.

$3^4 \cdot 9^2$

$3^4 \cdot (3^2)^2$

$3^4 \cdot (3^2 \cdot 3^2)$

3^8

We are going to keep our answers as a simplified version and not completely evaluate the expression at this time.

Law of exponents for division   The quotient of two numbers in exponential form with the same base is equal to that base with a power equal to the difference of the powers of each number.

To divide numbers with exponents that have the SAME base, subtract the exponents.

Examples:

$\dfrac{4^9}{4^4}$

Means $\dfrac{4 \times 4 \times 4 \times 4 \times 4 \times 4 \times 4 \times 4 \times 4}{4 \times 4 \times 4 \times 4}$

This is the same as 4^5.

You can get the same answer by subtracting the exponents of 9 and 4 to get the exponent of 5.

$\dfrac{8^8}{8^3}$

Means $\dfrac{8 \times 8 \times 8 \times 8 \times 8 \times 8 \times 8 \times 8}{8 \times 8 \times 8}$

This is the same as 8^5.

You can get the same answer by subtracting the exponents of 8 and 3 to get the exponent of 5.

To divide numbers with exponents that have a DIFFERENT base, try to make them the same base, if possible.

If you cannot make them into numbers with the same base, then you are just writing them together as one term.

$$\frac{3^9}{4^4} = 3^9 \times 4^{-4} \text{ or keep the same way}$$

If you can make them the same, first make the bases the same, then subtract the exponents.

$$\frac{2^4 \cdot 2^7}{4^2 \cdot 2^5}$$

$$\frac{2^{11}}{(2^2)^2 \cdot 2^5}$$

$$\frac{2^{11}}{2^4 \cdot 2^5}$$

$$\frac{2^{11}}{2^9}$$

$$2^2$$

We are going to keep our answers as a simplified version and not completely evaluate the expression.

EXAMPLE 8.4

Simplify.

1) $3^0 \cdot 3^7$

2) $8^1 \cdot 8^7$

3) $5^{-2} \cdot 3^7$

4) $9^3 \cdot 3^7$

5) $4^3 \cdot 2^7 \cdot 2^4$

6) $2^1 \cdot 3^7 \cdot 2^7 \cdot 4^2 \cdot 3^5 \cdot 3^9$

7) $\dfrac{3^0}{3^7}$

8) $\dfrac{8^{10}}{8^7}$

9) $\dfrac{9^3}{3^5}$

10) $\dfrac{2^4}{3^4 \cdot 2^5}$

11) $\dfrac{2^1 \cdot 3^7 \cdot 2^7}{4^2 \cdot 3^5 \cdot 3^9}$

SOLUTIONS

1) 3^7
2) 8^8
3) $5^{-2} \cdot 3^7$
4) 3^{13}
5) 2^{17}
6) $2^{12} \cdot 3^{21}$
7) 3^{-7}
8) 8^3
9) 3
10) $3^{-4} \cdot 2^{-1}$
11) $2^4 \cdot 3^{-7}$

8.5 What Is Scientific Notation?

First let's look back at two concepts.

The **standard form of a number** is the way a number normally looks using our base ten number system, with decimals and commas.

Examples:

2,400,000,000 or 0.000098

The **exponential form of a number** is the way to write the number using exponents. Since our number system is a base ten system, we can use powers of ten to rewrite numbers. This is generally done for really large or really small numbers using scientific notation.

Scientific notation   A form of writing a number as the product of a power of 10 and a decimal number greater than or equal to 1 and less than 10.

$$2,400,000 = 2.4 \times 10^6$$
$$240.2 = 2.402 \times 10^2$$
$$0.0024 = 2.4 \times 10^{-3}$$

If you were measuring something really small, you might use a nanometer. A nanometer is one-billionth of a meter (1 nanometer = .000000001 meter). Since a nanometer is so small and contains many zeroes, it is often more convenient to use scientific notation. If you were discussing a light year (9,460,528,400,000,000 meters)—something which is really large and contains many zeroes—it would be more convenient to use scientific notation.

To write a number in scientific notation:

First, write the number as a number between 1 and 10 by moving the decimal point to the appropriate place.

Multiply the new number by a power of ten equal to the number of places you moved the decimal point. The sign of the exponent will correspond to the power of ten needed to multiply. Large numbers will use positive exponents, since they are multiplied by powers of ten greater than 1; small numbers will use negative exponents, as they are multiplied by powers of ten less than 1.

Examples:

$$6,300,000,000$$
$$6.3 \times 10^9$$

$$0.000000456$$
$$4.56 \times 10^{-7}$$

The numbers are now much easier to compare. It is a clear, alternative way to writing numbers.

$$10^7 > 10^3 \qquad 10^{-2} > 10^{-6}$$

You can change the numbers back into standard form by reversing the process or multiplying.

1) Write the numbers in scientific notation.
   a) 34,000,000
   b) 100,000,000,000
   c) 0.0000000542
   d) 0.0000000000841

2) Write the numbers in standard form.
   a) 5.6×10^9
   b) 8.9×10^5
   c) 9×10^{-7}
   d) 1.32×10^{-9}

SOLUTIONS
1) a) 3.4×10^7
   b) 1×10^{11}
   c) 5.42×10^{-8}
   d) 8.41×10^{-11}

2) a) 5,600,000,000
   b) 890,000
   c) 0.0000009
   d) 0.00000000132

8.6 How Can We Evaluate Expressions with Exponents?

In Chapter 1, we learned that to evaluate numerical expressions with more than one operation, we need to use order of operations. We are going to try some more complicated examples to include higher exponents.

Examples:

$9 - 3^4 \times 2 + 7$	exponent of 4 (means $3 \times 3 \times 3 \times 3$)
$9 - 81 \times 2 + 7$	do the multiplication
$9 - 162 + 7$	do the subtraction
$-153 + 7$	do the addition
-146	

$12^2 \div 3 + 5^4 \times 8$	use the exponents
$144 \div 3 + 625 \times 8$	do the division
$48 + 625 \times 8$	do the multiplication
$4 + 5000$	do the addition
5004	

We learned, in Chapter 5, to evaluate algebraic expressions. We are going to try some more complicated examples to include higher exponents.

Examples:

Evaluate

$5x^2 + 6y^3 + 3x + 4y$ when $x = -2$ and $y = 5$
$5(-2)^2 + 6(5)^3 + 3(-2) + 4(5)$
$5(4) + 6(125) - 6 + 20$
$20 + 750 - 6 + 20$
784

Evaluate

$2x^2y^3 - 3xy^4$ when $x = 2$ and $y = -3$
$2(2)^2 (-3)^3 - 3(2)(-3)^4$
$2(4)(-27) - 3(2)(81)$
$-216 - 486$
-702

Chapter Review

1) Evaluate
   a) 2^3
   b) 2^{-4}
   c) 2^0
   d) $6^0 + 5 + 3^2$
   e) $9 - 3 \times 2^2 + 7$
   f) $4^2 \div 2 + 5 \times 8$
   g) $2^4 + 32 \div 8 - 6$

2) Write using exponents.
   a) $2 \cdot 2 \cdot 2 \cdot 2 \cdot 2 \cdot 2 \cdot 2 \cdot 2 \cdot 2$
   b) $12 \cdot 12 \cdot 12 \cdot 3 \cdot 3 \cdot 3$

3) Simplify
   a) $4^3 \cdot 2^7 \cdot 2^4$
   b) $2^4 \cdot 3^7 \cdot 2^5$
   c) $5^3 \cdot 3^7 \cdot 5^{-5}$
   d) $\dfrac{4^3}{2^7 \cdot 2^4}$
   e) $\dfrac{2^4}{3^4 \cdot 2^5}$
   f) $\dfrac{5^3 \cdot 3^7}{5^5}$

4) Write the numbers in scientific notation.
   a) 12,400,000,000
   b) .00000837

5) Write the numbers in standard form.
   a) 3.6×10^{-9}
   b) 5.4×10^6

6) Find the GCF:
   a) 16 and 56
   b) 75 and 135

7) Find the prime factorization of the following:
   a) 54
   b) 65

8) Find the LCM:
   a) 12 and 8
   b) 25 and 15

9) Evaluate
   $$15x^2 + 3y^3 + 2x - 4y \qquad \text{when } x = -3 \text{ and } y = 7$$

Solving Equations

WHAT YOU WILL LEARN

- How to understand solving an equation as a process of answering a question: which values from a specified set, if any, make the equation true.
- How to use inverse operations for solving one-variable equations.
- How to use substitution to determine whether a given number in a specified set makes an equation true.
- How to write equations to solve problems by reasoning about the quantities.
- How to solve word problems leading to equations.

SECTIONS IN THIS CHAPTER

- How Do We Solve a One-Step Equation?
- How Do We Solve a Two-Step Equation?
- How Can Combining Like Terms Help Us?
- How Can the Distributive Property Help Us?
- How Can We Solve Multi-Step Equations?
- How Can We Write an Equation to Solve Problems Algebraically?

Equation   A mathematical sentence stating that two expressions are equal.
Solution   The value or values that make an equation, inequality, or open sentence true.

9.1 How Do We Solve a One-Step Equation?

We are going to use tiles to represent equations.

The above model represents $x + 2 = 7$
x is represented by the rectangle.
2 is represented by the small squares on the left side of the equation.
7 is represented by the small squares on the right side of the equation.

We want to find the value of the x. We can tell by inspection the rectangle is worth 5 small squares. This can also be done by removing two small squares from each side.

The most important rule to remember with equations is keeping it equal. You have to "do the same thing to both sides."

Another way to look at it is to think of operations. The equation is showing addition. We want to "undo" the equation. We need the inverse of addition. We need subtraction. If we subtract 2 from both sides, we get the solution of $x = 5$ as well.

Let's look at another equation.

$$2x = -8$$

$2x$ is represented by the 2 rectangles.
-8 is represented by the small squares on the right side of the equation.

We want to find the value of the x. We can tell by inspection the rectangle is worth 4 small squares. This can also be done by removing equivalent amounts from both sides. If we take half of the left side, we remove one rectangle. If we take half of the right side, we take 4 squares.

Remember: the most important rule to remember with equations is keeping it equal. You have to "do the same thing to both sides."

Another way to look at it is to think of operations. The equation is showing multiplication. We want to "undo" the equation. We need the inverse of multiplication. We need division. If we divide by 2 on both sides, we get the solution of $x = -4$ as well.

In summary, since addition and subtraction are inverse operations, you can subtract to solve an addition equation and add to solve a subtraction equation. Since multiplication and division are inverse operations, you can divide to solve a multiplication equation and multiply to solve a division equation. Many times, you will be able to use your number sense to solve a one-step equation in your head.

Examples:

$$x - 3 = 14$$
$$+3 \qquad +3 \qquad \text{add 3 to both sides}$$
$$x = 17$$

Check to make sure you are correct.
Substitute the found value for the variable

$$(17) - 3 = 14$$
$$14 = 14$$

The equation is true; your solution is correct.

Another way you can solve the equation is to use guess and check. You can probably guess of a number that, when you subtract 3, will have a resulting difference of 14.

$$\text{Try:} \quad 18 - 3 = 15$$

$$\text{Try again:} \quad 17 - 3 = 14$$

Therefore, x will equal 17.

$$\frac{x}{5} = -7$$
$$5\left(\frac{x}{5}\right) = (-7)5 \qquad \text{multiply both sides by 5}$$
$$x = -35$$

Check to make sure you are correct.
Substitute the found value for the variable

$$\frac{(-35)}{5} = -7$$
$$-7 = -7$$

The equation is true; your solution is correct.

You can probably guess of a number that, when you divide by 5, will have a resulting quotient of -7.

$$\text{Try:} \quad \frac{-30}{5} = -6$$
$$\text{Try again:} \quad \frac{(-35)}{5} = -7$$

Therefore, x will equal -7.

EXAMPLE 9.1

1) $x + 4 = -2$
2) $3 + x = 7$
3) $x - 9 = 11$
4) $2x = -2$
5) $\frac{y}{4} = 7$
6) $42 = 7x$

SOLUTIONS

1) -6
2) 4
3) 20
4) -1
5) 28
6) 6

9.2 How Do We Solve a Two-Step Equation?

A two-step equation involves two operations. You are going to need two inverse operations to solve it. Since solving an equation is like "un-doing" a problem, we use the reverse idea of order of operations. We would generally do any adding or subtracting before multiplying or dividing. Always use the inverse operation of what you see in the equation. Think about going in reverse: do the inverse.

$$
\begin{aligned}
5x + 2 &= -8 \\
-2 \quad &\quad -2 \qquad \text{subtract 2 from both sides} \\
\frac{5x}{5} &= \frac{-10}{5} \qquad \text{divide both sides by 5} \\
x &= -2
\end{aligned}
$$

Check to make sure you are correct.
Substitute the found value for the variable
$$
\begin{aligned}
5(-2) + 2 &= -8 \\
-10 + 2 &= -8 \\
-8 &= -8
\end{aligned}
$$

The equation is true; your solution is correct.

Add two reds to each side

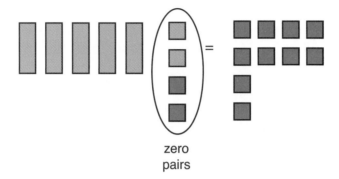

zero
pairs

Divide into 5 groups

Each group is worth -2.

$$\frac{y}{7} - 1 = 12$$
$$+1 = +1 \qquad \text{add 1 to both sides}$$
$$7\left(\frac{y}{7}\right) = (13)\, 7 \qquad \text{multiply both sides by 7}$$
$$x = 91$$

Check to make sure you are correct.
Substitute the found value for the variable

$$\frac{(91)}{7} - 1 = 12$$
$$13 - 1 = 12$$
$$12 = 12$$

The equation is true; your solution is correct.

$$-10 = 3x - 1 \qquad \text{both sides are equal, you can reflect it if you want}$$
$$+1 = +1 \qquad \text{add 1 to both sides}$$
$$\frac{-9}{3} = \frac{3x}{3} \qquad \text{divide both sides by 3}$$
$$-3 = x$$
$$x = -3 \qquad \text{Answer with the variable first}$$

Check to make sure you are correct.
Substitute the found value for the variable

$$-10 = 3(-3) - 1$$
$$-10 = -9 - 1$$
$$-10 = -10$$

The equation is true; your solution is correct.

EXAMPLE 9.2   Solve the following equations.
1) $3x - 8 = -8$
2) $6 = 4x - 6$
3) $-8 + 3x = -5$
4) $-5c + 8 = 18$
5) $\frac{x}{6} + 2 = -4$

SOLUTIONS
1) $x = 0$
2) $x = 3$
3) $x = 1$
4) $c = -2$
5) $x = -36$

9.3 How Can Combining Like Terms Help Us?

Think about your room being a mess. There are shoes all around. It is easier to find things when they are together. Like terms work the same way. You just need to get them together. We should simplify the equation and then solve it.

Simplify first $2x + 4x = 12$

$$\frac{6x}{6} = \frac{12}{6}$$ Combine the like terms

$$x = 2$$

Each group is worth 2.

It's much easier when simplified!!!

Simplify first

$$4y + 3y - 5 = 16$$
$$7y - 5 = 16$$
$$+5 = +5$$
$$\frac{7y}{7} = \frac{21}{7}$$
$$y = 3$$

Check to make sure you are correct.
Substitute the found value for the variable

$$4(3) + 3(3) - 5 = 16$$
$$12 + 9 - 5 = 16$$
$$16 = 16$$

The equation is true; your solution is correct.

Simplify first

$$18 = 5y + 6 - 8y + 3$$
$$18 = -3y + 9$$
$$-9 = -9 \qquad \qquad \text{subtract 9 from both sides}$$
$$\frac{9}{-3} = \frac{-3y}{-3} \qquad \qquad \text{divide both sides by } -3$$
$$-3 = y \qquad \qquad \text{reflect}$$
$$y = -3$$

Check to make sure you are correct.
Substitute the found value for the variable

$$18 = 5(-3) + 6 - 8(-3) + 3$$
$$18 = -15 + 6 + 24 + 3$$
$$18 = 18$$

The equation is true; your solution is correct.

1) $3x - 3 + 2x = 12$
2) $12y + 2y = -28$
3) $-4x + 2x = -20$
4) $12 = 4x + 3 - 2x$
5) $5x - 7 + 12 = -5$

SOLUTIONS

1) $x = 3$
2) $y = -2$
3) $x = 10$
4) $x = 4.5$
5) $x = -2$

9.4 How Can the Distributive Property Help Us?

First we will practice the distributive property.

$$\frac{1}{2}(4x - 8)$$

$$2x - 4$$

$$3(2x + 5) = 6x + 15$$
$$-4(x - 7) = -4x + 28$$
$$7(2x + 3) = 14x + 21$$

When solving an equation which has the distributive property, handle the distribution first. Once you have distributed, you have a two-step equation which you can solve.

Distribute first
$$3(x + 2) = 15$$
$$3x + 6 = 15$$
$$\underline{-6 \quad -6}$$
$$\frac{3x}{3} = \frac{9}{3}$$
$$x = 3$$

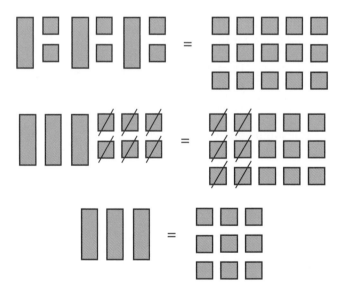

Each group is worth 3.

Distribute first
$$4(3x - 9) = 24$$
$$12x - 36 = 24$$
$$\underline{+36 \qquad +36}$$
$$\frac{12x}{12} = \frac{60}{12}$$
$$x = 5$$

EXAMPLE 9.4

Solve for the given variable:

1) $4(y - 2) = 20$
2) $12(x - 2) = -48$
3) $\frac{1}{2}(16y + 6) = 19$
4) $-21 = -3(x - 5)$
5) $\frac{3}{4}(4x + 12) = 18$
6) $2x + 3(4x + 5) = 1$

SOLUTIONS

1) $y = 7$
2) $x = -2$
3) $y = 2$
4) $x = 12$
5) $x = 3$
6) $x = -1$

9.5 How Can We Solve Multi-Step Equations?

Your move will depend on the left expression and right expression. Draw the line; simplify each side and decide before you move. Make your decision based on your comfort level. Don't be afraid of negative numbers. As long as you calculate correctly, there is no bad decision. Remember to look before you leap!

$5b + 7 = 3b - 17$	both sides are simplified
$ -3b -3b$	combine the like terms by moving to the same side of the equation
$2b + 7 = -17$	subtract 7 from both sides
$ -7 -7$	
$\frac{2b}{2} = \frac{-24}{2}$	divide both sides by 2
$b = -12$	

Check to make sure you are correct.
Substitute the found value for the variable

$$5(-12) + 7 = 3(-12) - 17$$
$$-60 + 7 = -36 - 17$$
$$-53 = -53$$

The equation is true; your solution is correct.

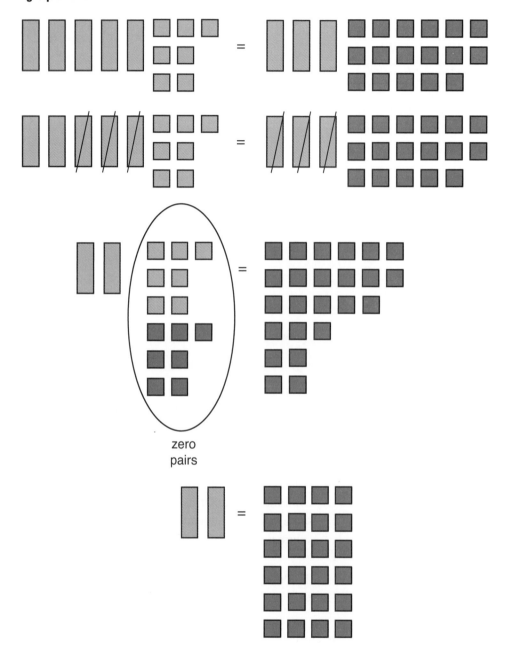

zero
pairs

Each group is worth -12.

$$8 + 9p - 4 = 6p - 2 \qquad \text{simplify the right side}$$
$$9p + 4 = 6p - 2 \qquad \text{get all the variables on one side}$$
$$-6p \qquad = \qquad -6p \qquad \text{subtract } 6p \text{ from both sides}$$
$$3p + 4 = -2$$
$$-4 = -4 \qquad \text{subtract 4 from both sides}$$
$$\frac{3p}{3} = \frac{-6}{3} \qquad \text{divide both sides by 3}$$
$$p = -2$$

$$5 - 6x = 2x + 21$$
$$-2x = -2x$$
$$-8x + 5 = 21$$
$$-5 = -5$$
$$\frac{-8x}{-8} = \frac{16}{-8}$$
$$x = -2$$

EXAMPLE 9.5

Solve for the given variable:

1) $x - 5 = 4x + 19$
2) $4x - 5 = 2x + 17$
3) $-4x + 8 = 2x - 16$
4) $16 + 2x = 4x - 8$
5) $-3x + 3 = 2x + 8$

SOLUTIONS

1) $x = -8$
2) $x = 11$
3) $x = 4$
4) $x = 12$
5) $x = -1$

9.6 How Can We Write an Equation to Solve Problems Algebraically?

Write an equation to solve each of the following problems. Be sure to assign a variable and answer the question.

Assigning a variable

You can use any lowercase letter. You may want to draw a picture to help or associate the letter with the story.

Write the equation

Use the words of the story to write an equation.

Solve your equation

Follow all the steps needed. Be careful of integer rules.

Check your answer

Don't check the equation. Check your answer with the words of the story. Use your best judgment and common sense.

Answering the question

Write an "answer blank sentence" to help focus your attention.

Examples:

John and Pat are twin brothers. The sum of their ages is 32. How old is John?

Assignment of variables: Let j = John's age

Let j = Pat's age (they are twins)

Equation: $j + j = 32$

Solution: $j = 16$

Check: $16 + 16 = 32$

Answer: John is 16 years old.

Ana and Chris ran a total of 15 miles. Chris ran 3 more miles than Ana. How many miles did they each run?

Assignment of variables: Let a = Ana's miles

Let $a + 3$ = Chris's miles (3 more than Ana)

Equation: $a + a + 3 = 15$

Solution: $2a + 3 = 15$

$2a = 12$

$a = 6$

Check: $6 + 6 + 3 = 15$

Answer: Ana ran 6 miles and Chris ran 9 miles.

EXAMPLE 9.6

1) When 10 is added to half the sum of a number and 8, the result is 30. What is the number?
2) Jose bought 3 CDs (each for the same price), and a set of earphones for a price of $10. The total cost before tax was $49. What was the cost of one CD?
3) Nick's age plus his sister Tanya's age is 48 years old. Tanya's age is half Nick's age plus 6 years. How old is Tanya? How old is Nick?
4) Tickets for a Ducks' game for one adult and two children cost a total of $21. The adult ticket is $3 more than a child's ticket. Find the cost of each ticket.
5) The product of 8 and the difference between a number and 9 is 96. What is the number?

SOLUTIONS

1) The number is 32.
2) Each CD was $13.
3) Nick is 28.
4) The children's tickets are $6 and the adult ticket is $9.
5) The number is 21.

Cumulative Review

The following two problems have been solved INCORRECTLY. Find the errors and **re-do** *the problem correctly.*

1. $-4 + 3x = -13$

$$\underline{+4 \qquad +4}$$
$$8 + 3x = -9$$
$$\underline{-8 \qquad -8}$$
$$\frac{3x}{3} = \frac{-1}{3}$$

$$x = -0.3$$

2. $3 - 6a = 15$

$$\underline{+3 \quad +3}$$
$$\frac{6a}{6} = \frac{18}{6}$$

$$a = 3$$

Solve each equation.

3) $3x + 4 = 10$

4) $5(b + 2) = -45$

5) $5x - 3 = 7$

6) $15 - 6x = 4x + 25$

7) $8 + 9p = 6p - 22$

8) $27 = 3y + 16 - 6y + 2$

9) $24 = 3y - 9y$

10) $-2(p - 12) = 24$

11) Solve algebraically: The square and rectangle below have the same perimeter. Write an equation to find the lengths of the sides of the rectangle. All the measurements are in centimeters. Find the sides of the rectangle.

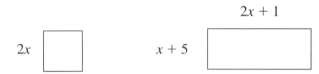

12) A student measured his textbook to be 34 cm when the ACTUAL value was really 30 cm. What is the student's percent of error to the nearest tenth?

13) Ashley bought her shirt from Armani Exchange for $65.00. Alex bought the same shirt three weeks later on sale. Alex paid $39.00. What percent did Alex save?

14) Nickel is made of silver and aluminum. If the ratio of silver to aluminum is 1 to 10, how many pounds of silver are there in 66 pounds of nickel?

15) Add $\frac{3}{5} + \frac{7}{12}$

16) Find the prime factorization of 56.

17) List the first 7 multiples of 11.

18) Find the unit price when you purchase 6 bananas for a total of $2.94.

19) Evaluate $x^3 - 2x + 7$ when $x = -2$

20) Solve $\frac{4}{x} = \frac{52}{143}$

Solving Inequalities

WHAT YOU WILL LEARN

- How to write an inequality to represent a constraint or condition in a real-world or mathematical problem.
- How to recognize that inequalities have infinitely many solutions; represent solutions of such inequalities on number line diagrams.
- How to solve inequalities and represent the solution set.
- How to write inequalities to solve problems by reasoning about the quantities.
- How to solve word problems leading to inequalities.
- How to graph the solution set of the inequality and interpret it in the context of the problem.

SECTIONS IN THIS CHAPTER

- What Is an Inequality?
- How Do We Represent Solutions of Inequalities?
- How Do We Solve Inequalities?
- How Can We Use Inequalities to Solve Word Problems?

10.1 What Is an Inequality?

Inequality A mathematical statement containing one of the symbols $>$, $<$, $\geq$, $\leq$, or $\neq$ to indicate the relationship between two quantities.

An inequality tells you when things are not equal. There are many times when an exact number isn't needed. Think about situations when you are given a minimum or a maximum. Those situations generate inequalities.

For example:

A curfew of 11 p.m. means be home at or before 11 p.m.

> You can come home at 9 p.m., 10 p.m., or 10:30 p.m.—even 11 p.m.
> You'd better not come home at 11:30; you would be in trouble.

You have to be at least 18 years old to vote.

> You can be 18, 19, 25, or even 75 (like my Uncle Carl).
> You can't be 16, 12, or even 4 (like Charlie).

You can be at most 12 years to order from the kid's menu.

> You can be 12, 11, or even just a few months (like Luke).
> You can't be 13, 16, or 26 (like Chris).

A party room has a maximum capacity of 125 people.

> You can have 125, 124, or only 6.
> You can't have 126, 130, or 250. The fire marshal will shut it down.

All of the above situations can be written as an inequality using the following symbols.

Symbol	Meaning
$>$	Is greater than
$<$	Is less than
$\geq$	Is greater than or equal to; is at least
$\leq$	Is less than or equal to; is at most
$\neq$	Is not equal to

$x > 4$ a number is greater than 4

$x < 4$ a number is less than 4

$x \geq 4$ a number is greater than or equal to 4

$x \leq 4$ a number is less than or equal to 4

$x \neq 4$ a number does not equal 4

Generally in an algebraic inequality, the symbol is written first. If it is not, then use the reflexive property to "turn it around."

$12 > x$ would become $x < 12$ keeping the relationship the same; look at how x remained the lower number.

EXAMPLE 10.1

Write the inequality to represent the given situation.
1) A number is at most -2.
2) A number is less than 12.
3) A number is at least 17.
4) A number is greater than $\frac{1}{2}$.
5) You have to be at least 54 inches to ride the roller coaster.
6) You have only $50 to spend on vacation.

SOLUTIONS
1) $x \leq -2$
2) $x < 12$
3) $x \geq 17$
4) $x > \frac{1}{2}$
5) $x \geq 54$
6) $x \leq 50$

10.2 How Do We Represent Solutions of Inequalities?

Solution set   The set of values that make an equation or statement true.

A solution set can be shown on a number line or as a set.

Example: $x > 6$

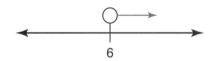

6

To make a graph:
Make a circle at the number in the inequality
If you are including the number because you have a $\leq$ or $\geq$ symbol, fill the circle in.
If you are not including the number because you have a $<$ or $>$ symbol (no equal to), leave the circle open.

Shade the number line.
If you have a $\leq$ or $<$ symbol; shade to the left ("less than" means go left).
If you have a $\geq$ or $>$ symbol; shade to the right ("greater than" means go right).

More examples:

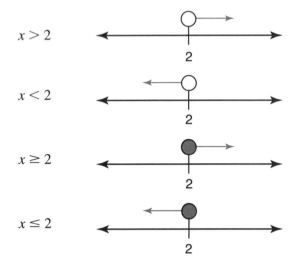

$x > 2$

$x < 2$

$x \geq 2$

$x \leq 2$

The number line is a good illustration. It is impossible to list all the numbers because we are naming all real numbers. Real numbers are dense. You can always name another one in between the two stated. Also, we are going to infinity or negative infinity if less than.

You can use set notation but these would only be used for integers not the set of real numbers.

To use set notation:

Start with braces. Decide whether the given number is part of the set or just a border. If it is to be included, write it down. If it is not to be included, start with the next available number.

Examples:

$x > 2$	$\{3,4,5,6,\ldots\}$	doesn't include the 2 and goes up
$x < 2$	$\{1,0,-1,-2,\ldots\}$	doesn't include the 2 and goes down
$x \geq 2$	$\{2,3,4,5,6,\ldots\}$	includes the 2 and goes up
$x \leq 2$	$\{2,1,0,-1,-2,\ldots\}$	includes the 2 and goes down

EXAMPLE 10.2 Graph the inequalities and write the solution set in set notation for integers.

1) $x < 3$
2) $x > -4$
3) $2 \leq x$
4) $x > 7$
5) $x \leq -1$

SOLUTIONS

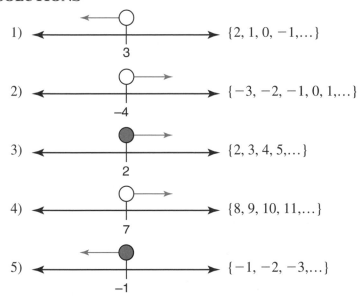

1) {2, 1, 0, −1,...}
2) {−3, −2, −1, 0, 1,...}
3) {2, 3, 4, 5,...}
4) {8, 9, 10, 11,...}
5) {−1, −2, −3,...}

10.3 How Do We Solve Inequalities?

Solving an inequality is almost exactly the same as solving an equation.

Simplify each side of the inequality
Combine all the like terms
Add or subtract
Multiply or divide * this is where you need to be careful. Here is the change:

IMPORTANT NOTE

If you divide by or multiply by a negative number, you need to reverse the direction of the inequality symbol.

Here is an example of reversal of signs when dividing by a negative:

$$-2x - 3 > 9$$
$$ +3 \quad +3 \qquad \text{add 3 to both sides}$$
$$\frac{-2x}{-2} > \frac{12}{-2} \qquad \text{divide both sides by } -2$$

What happens when you divide by a negative using integers?
Positive ÷ negative = negative ($12 \div -2 = -6$)
Negative ÷ negative = positive ($-2 \div -2 = 1$)
It changes the sign!
Same idea: a $>$ becomes a $<$

$x < -6$

You would then graph the solution set.

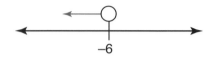

Another example:

$$8x - 4x + 12 < 2x + 18$$

$4x + 12 < 2x + 18$	Simplify the left side
$-2x \qquad -2x$	Combine the like terms
$2x + 12 < 18$	
$-12 \quad -12$	Subtract 12 from both sides
$\dfrac{2x}{2} < \dfrac{6}{2}$	Divide both sides by 2

$$x < 3$$

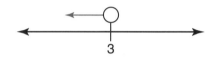

EXAMPLE 10.3

Solve and graph.

1) $4(x - 2) > 8$
2) $\frac{1}{2}x + 7 \le 5$
3) $6x + 12 > -4x - 8$
4) $5x + 9 - 8x \ge -9$
5) $3x + 7 < -2$

SOLUTIONS

1) $x > 4$

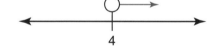

2) $x \le -4$

3) $x > -2$

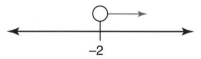

4) $x \le 6$

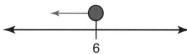

5) $x < 3$

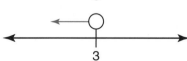

10.4 How Can We Use Inequalities to Solve Word Problems?

You are going to read the story, pull out important information, write the inequality, solve the problem, and answer the questions.

Remember the five steps used to solve problems with equations?

- Assigning a variable
- Write the equation
- Solve your equation
- Check your answer
- Answering the question

What should we change about these five steps?

- Assigning a variable
- Write the inequality
- Solve your inequality
- Check your answer
- Answering the question: this is a little more difficult because we have a solution set, not a single solution.

Example:

1) The shallow end of the pool is less than one-quarter as deep as the deep end of the pool. The shallow end is 3 feet deep. Write an inequality to solve this problem.

   Assign the variable: $d =$ deep end of pool

   Write the inequality: $3 < \frac{1}{4}d$

   Solve the inequality: $4(3) < (\frac{1}{4}d)\,4$

   $12 < d$ therefore $d > 12$

   Check: try a few values. If it is 16, then the shallow end needs to be less than 4 feet. 3<4 it works!

   Answer: The deep end is greater than 12 feet deep.

2) In order to have the $200 he needs for a bike, Jack plans to put money away each week for the next 15 weeks. What is the minimum amount in dollars that Jack will need to save on average each week in order to reach his goal?

   Assign the variable: $d =$ dollars Jack needs to save per week

   Write the inequality: $15d \geq 200$

Think: you need at least 200, you can't go to the store with less. But it is fine to bring more.

Solve the inequality: $\dfrac{15d}{15} \geq \dfrac{200}{15}$

$d \geq 13.33$

Check: try a few values. If it is $13.33, then he will have $199.95. We need to go higher. Also, the question asks for the amount in whole dollars. We need to try $14; if it is $14, then he will have $210 in 15 weeks, which will work.

Answer: He will need to save a minimum of $14 per week.

3) A t-shirt retailer must pay $120 for a design and $4 per shirt. How many t-shirts would he have to sell at $9 per shirt to make a profit? Remember, to make a profit your RECEIPTS must be greater than your COSTS.

Assign the variable: $x =$ the number of t-shirts

Write the inequality: $9x > 4x + 120$

Solve the inequality: $9x > 4x + 120$

$5x > 120$

$x > 24$

Check: We need to try 24; if it is even or less, then we know 25 is correct.

Answer: 25 t-shirts

EXAMPLE 10.4

1) The class representatives are planning a dance. The food, decorations, and entertainment cost $350. They already have $75 saved for the dance. If they sell tickets at $5 each, at least how many tickets must be sold to cover the remaining cost of the dance?

2) At laser tag, the party price for the first 15 kids is $229 and $12 for each additional kid. What is the greatest number of kids who can be at the party for less than $350.00? (be careful when you answer)

3) Eddie's elementary school is having a fall carnival. Admission to the carnival is $3, and each game inside the carnival costs $.25. Write an inequality that represents the possible number of games that can be played having $10. What is the maximum number of games that can be played?

4) You and your family attend your brother's championship baseball game. Between innings, you decide to go to the snack stand. You go to the snack stand with $15 and find that sodas are $2.50 and popcorn is $1.75. Write and solve an inequality that can be used to find the maximum number of sodas if you buy 4 bags of popcorn.

5) Charlene wants to organize 127 CDs into storage boxes. Each storage box can hold a maximum of 10 CDs. What is the minimum number of storage boxes needed?

SOLUTIONS

1) $5x + 75 \geq 350$ $x \geq 55$ They must sell 55 tickets.
2) $12x + 229 < 350$ $x < 10.1$ You can invite 10 additional kids for a total of 25.
3) $3 + .25x \leq 10$ $x \leq 28$ You can play at most 28 games.
4) $2.50x + 4(1.75) \leq 15$ $x \leq 3.2$ You can buy at most 3 sodas.
5) $10x \geq 127$ $x \geq 12.7$ Charlene needs 13 boxes.

Chapter Review

1) Which of the following values is in the solution set of the inequality?
$5x + 3 > 38$
(1) 5   (3) 7
(2) 6   (4) 8

2) Using the set of positive integers, what is the solution set of the inequality
$2x - 3 < 6$?
(1) {0,1, 2, 3}   (3) {0,1, 2, 3, 4}
(2) {1, 2, 3}   (4) {1, 2, 3, 4}

Graph the inequalities.

3) $x \leq 8$

4) $x \geq 5$

5) $0 < x$

6) $x \leq 6$

Solve and graph.

7) $-2x + 7 > 15$

8) $3x - 6 + 2x < 4$

9) $4(x + 3) \leq -12$

10) As a salesperson, you are paid $50 per week plus $3 per sale. This week you want your pay to be at least $100. Write an inequality for the number of sales you need to make, and describe the solutions.

11) In order to have the $2,000 he needs for a car, Charlie plans to put money away each week for the next 35 weeks. What is the minimum amount in dollars that Charlie will need to average each week in order to reach his goal?

12) A skirt retailer must pay $220 for beads and $40 per jean skirt. How many skirts would she have to sell at $75 per jean skirt to make a profit? (Remember, to make a profit, your RECEIPTS must be greater than your COSTS.)

Geometry

WHAT YOU WILL LEARN

- To use facts about supplementary, complementary, vertical, and adjacent angles to write and solve simple equations for an unknown angle.
- To use facts about supplementary, vertical, and adjacent angles to write and solve simple equations for an unknown angle when parallel lines are cut by a transversal.
- To recognize and classify two-dimensional figures.
- To recognize, understand, and calculate perimeter and area.
- To recognize, understand, and calculate area and circumference of a circle.
- To recognize and classify three-dimensional figures.
- To recognize, understand, and calculate surface area and volume.
- To describe the two-dimensional figures that result from slicing three-dimensional figures.

SECTIONS IN THIS CHAPTER

- What Are Angle Pair Relationships?
- What Are Vertical Angle Pairs?
- What Happens When Parallel Lines Are Cut by a Transversal?
- How Do We Classify Two-Dimensional Figures?
- What Is Area and Perimeter and How Do We Calculate Them?
- How Do We Classify Three-Dimensional Figures?
- What Is Surface Area and Volume and How Do We Calculate Them?

11.1 What Are Angle Pair Relationships?

First, let's review angles.

Angle　a figure formed by two rays with a common endpoint

Angles are classified by their angle measure. You measure an angle with a protractor. The measure is found in degrees. There are five classifications.

Acute—less than 90 degrees
Right—exactly 90 degrees
Obtuse—between 90 and 180 degrees
Straight—exactly 180 degrees
Reflex—between 180 and 360 degrees

We are going to look at pairs of angles which have specific relationships.

> **Complementary angles**—two angles whose sum is 90 degrees

Angle	The Complement
20°	70°
45°	45°
62°	28°
80°	10°
65°	25°

Notice that if you add the angles together, you will have a sum of 90 degrees. This can also be seen as a right angle if the angles are adjacent.

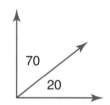

Supplementary angles—two angles whose sum is 180 degrees

Angle	The Supplement
120°	60°
45°	135°
62°	118°
110°	70°
65°	115°

Notice also that if you add the angles together, you will have a sum of 180 degrees. This can also be seen as a line or linear pair if the angles are adjacent.

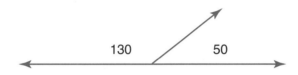

Congruent angles—angles that have the same measure

Angle	Congruent Angle
120°	120°
45°	45°
62°	62°

Notice the angles have the same measure. This can also be seen when an angle is bisected. An angle bisector is a line or ray which divides the angle into two congruent parts.

We can find the missing angle by simple subtraction or by writing a one-step equation.

Examples: Find the complement of 30 degrees.

$$90 - 30 = 60 \qquad \text{subtraction}$$
$$x + 30 = 90 \qquad \text{one-step equation}$$
$$x = 60$$

Find the supplement of 40 degrees.

$$180 - 40 = 140 \qquad \text{subtraction}$$
$$x + 40 = 180 \qquad \text{one-step equation}$$
$$x = 140$$

Sometimes the questions involve more than one step. This is usually when algebra is best.

Examples:

Two angles are complementary. One angle is twice the other angle. Find the angles.

Let x = the first angle
Let $2x$ = the second angle (because it is twice the other angle)

$x + 2x = 90$	they are complementary
$3x = 90$	combine like terms
$x = 30$	divide by the coefficient of 3

The first angle is 30 degrees; the second is 60 degrees ($2(30) = 60$).

Two supplementary angles are in a ratio of 7:3. Find the angles.

Let $7x$ = the first angle
Let $3x$ = the second angle

$7x + 3x = 180$	they are supplementary
$10x = 180$	combine the like terms
$x = 18$	divide by the coefficient of 10

The first angle is 126 degrees ($7(18) = 126$). The second angle is 54 degrees ($3(18) = 54$).

1. Find the complement of 18 degrees.
2. Find the supplement of 60 degrees.
3. Name an angle pair which is both congruent and complementary.
4. Two angles are complementary. One angle is ten more than four times the other angle. Find the angles.
5. Two angles are complementary. One angle is five less than four times the other angle. Find the angles.
6. Two angles are supplementary. One angle is ten less than nine times the other angle. Find the angles.

SOLUTIONS

1) 72 degrees
2) 120 degrees
3) 45 and 45
4) 16 and 74
5) 19 and 71
6) 19 and 161

11.2 What Are Vertical Angles?

First, let's review linear pairs. A linear pair consists of two angles which form a line. The linear pairs below are 1 and 2, 2 and 3, 3 and 4, and 4 and 1. The adjacent angles are linear pairs. It all depends on which line you are looking at to determine the pair.

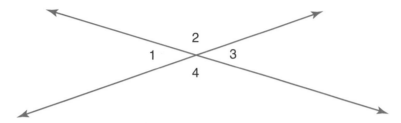

We know linear pairs are supplementary. Therefore, if we know one of the linear pair, with a little bit of subtraction we can find the other (Buy one, get one free!).

If we know Angle 1 is 70 degrees, we can find Angle 2 and Angle 4 by subtracting from 180. We now know Angle 1 is 70 degrees, Angle 2 is 110 degrees and Angle 4 is 110 degrees. This means that Angle 3 must also be 70 degrees because Angle 3 forms a linear pair with both Angles 2 and 4. This makes Angle 1 and Angle 3 congruent. This gives us a new relationship between the non-adjacent angles.

Vertical Angles: Two angles are vertical angles if their sides form two pairs of opposite rays. They are the non-adjacent angles when two lines intersect.

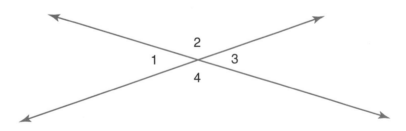

Now look at the VERTICAL ANGLES.
The vertical angle pairs are 1 and 3, and 2 and 4.

We can relate every angle to the other angles by using the words **congruent** or **supplemental**.

For example,

Angle 1 is supplemental to Angle 2.
Angle 1 is congruent to Angle 3.
Angle 1 is supplemental to Angle 4.

We can use this information to find the missing angles. **BUY ONE, GET THREE FREE!**

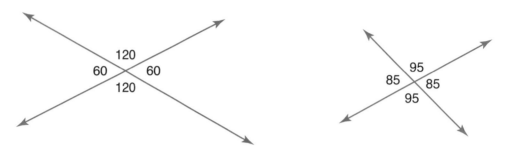

Remember halfway around is 180 degrees, and all the way around is 360 degrees. Vertical angles are congruent. Adjacent angles are supplemental.

To use algebra, you must first define the relationship, then set up the equation.

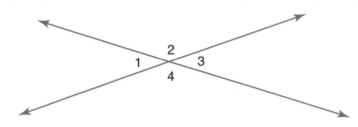

We can relate every angle to the other angles by using the words **congruent** or **supplemental**.

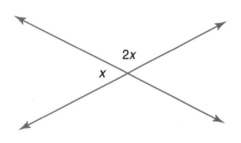

Relationship: Supplemental

Equation: $x + 2x = 180$
$3x = 180$
$x = 60$

Angles: 60
$2(60) = 120$

Relationship: Congruent

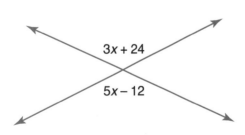

Equation: $3x + 24 = 5x - 12$
$24 = 2x - 12$
$36 = 2x$
$x = 18$

Angles: $3(18) + 24 = 78$
$5(18) - 12 = 78$

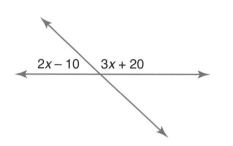

Relationship: Supplementary

Equation: $2x - 10 + 3x + 20 = 180$
$5x + 10 = 180$
$5x = 170$
$x = 34$

Angles: $2(34) - 10 = 58$
$3(34) + 20 = 122$

EXAMPLE 11.2

1) Find the missing angles.

a)

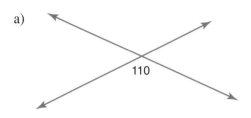

b)

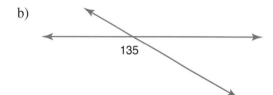

c)

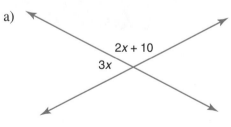

d)

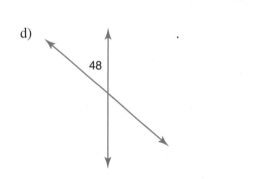

2) Define the relationship. Write the equation. Solve the equation. Find the missing angles.

a)

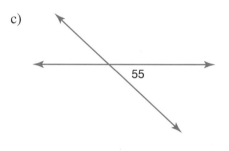

b)

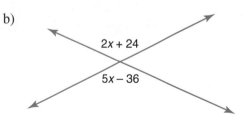

c)

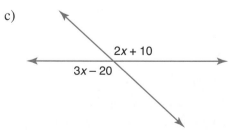

d)

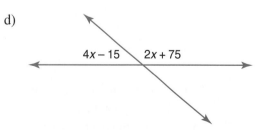

SOLUTIONS

1) a)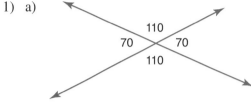

    b)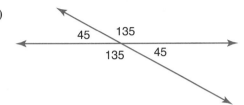

    c)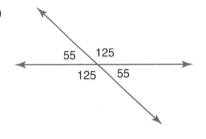

    d)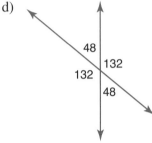

2) a) Supplementary
        $3x + 2x + 10 = 180$
        $x = 34$
        Angles: 102 and 78

    b) Congruent
        $5x - 36 = 2x + 24$
        $x = 20$
        Angles: 64 and 116

    c) Congruent
        $3x - 20 = 2x + 10$
        $x = 30$
        Angles: 70 and 110

    d) Supplementary
        $4x - 15 + 2x + 75 = 180$
        $x = 20$
        Angles: 65 and 115

11.3 What Happens When Parallel Lines Are Cut by a Transversal?

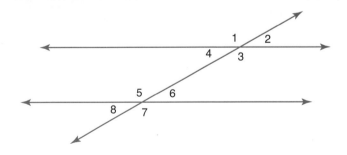

We know from our prior section all about Angles 1, 2, 3, and 4.

Angle 1 is congruent ($\cong$) to Angle 3. Angle 2 is congruent to Angle 4.

Angle 1 is supplementary to both Angles 2 and 4. The same is true of Angle 3.

Given the two horizontal lines are parallel; we can relate all the angles to each other in the same manner.

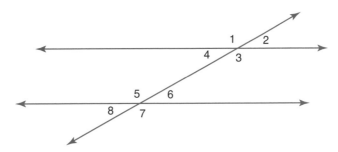

When parallel lines are cut by a transversal, **two identical sets** of vertical angles are formed.

The congruent angles have special names for their relationships.

Corresponding angles are in the same relative position. If you slide the top set of vertical angles down the transversal, the corresponding angles would match up.

$$\angle 1 \cong \angle 5$$
$$\angle 2 \cong \angle 6$$
$$\angle 3 \cong \angle 7$$
$$\angle 4 \cong \angle 8$$

Alternate Interior angles are between the two parallel lines on opposite sides of the transversal.

$$\angle 3 \cong \angle 5$$
$$\angle 4 \cong \angle 6$$

Alternate Exterior angles are outside the two parallel lines on opposite sides of the transversal.

$$\angle 1 \cong \angle 7$$
$$\angle 2 \cong \angle 8$$

Knowing these relationships and linear pairs, we can find all the angles.

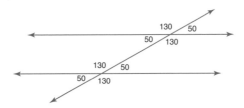

One angle is usually given and you need to find the other seven angles. Generally, start with all the congruent angles by following the pattern.

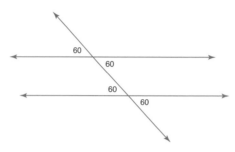

You can find the remaining angles by subtracting from 180. The remaining angles are supplemental to the given angle and congruent to each other.

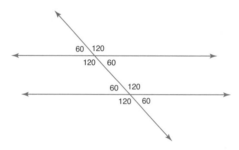

You can use algebra to find the missing angles as well. Follow the same idea as with vertical angles.

Define the relationship.
Set up the equation.
Solve your equation.
Find the angles.

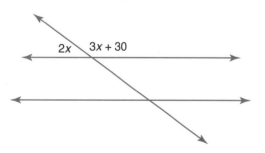

Relationship: Supplemental
Equation: $2x + 3x + 30 = 180$
                $5x + 30 = 180$
                $5x = 150$
                $x = 30$

Angles: $2(30) = 60$
                $3(30) + 30 = 120$

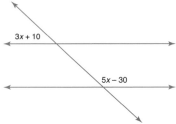

Relationship: Supplemental

Equation: $3x + 10 + 5x - 30 = 180$
$$8x - 20 = 180$$
$$8x = 200$$
$$x = 25$$

Angles: $2(25) + 10 = 60$
$$5(25) - 30 = 120$$

Relationship: Congruent

Equation: $3x - 30 = x + 10$
$$2x - 30 = 10$$
$$2x = 40$$
$$x = 20$$

Angles: $(20) + 10 = 30$
$$3(20) - 30 = 30$$

The supplementary angles are 150.
$$180 - 30 = 150$$

 EXAMPLE 11.3   Use the diagram below for questions 1–10.

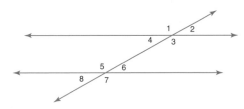

Name the relationship between the two angles and the reason for the relationship.

1) Angle 1 and Angle 2
2) Angle 3 and Angle 5
3) Angle 4 and Angle 8
4) Angle 6 and Angle 7
5) Angle 2 and Angle 8

Given one angle, find the other angle.

6) If Angle 6 is 130 degrees, what is the measure of Angle 7?
7) If Angle 5 is 30 degrees, what is the measure of Angle 7?
8) If Angle 3 is 70 degrees, what is the measure of Angle 8?
9) If Angle 1 is 65 degrees, what is the measure of Angle 8?
10) If Angle 2 is 80 degrees, what is the measure of Angle 4?

Define the relationship. Write the equation. Solve the equation. Find the missing angles.

11)

12)

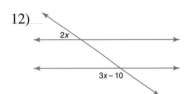

13)

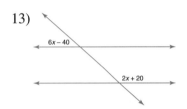

SOLUTIONS

    1) supplementary   linear pair

    2) congruent   alternate interior angles

    3) congruent   corresponding angles

    4) supplementary   linear pair

    5) congruent   alternate exterior

    6) 50 degrees

    7) 30 degrees

    8) 110 degrees

    9) 115 degrees

  10) 80 degrees

  11) congruent

$$4x - 30 = 2x + 38$$

$$x = 34$$

Angles: 106 and 74

  12) supplementary

$$2x + 3x - 10 = 180$$

$$x = 38$$

Angles: 76 and 104

  13) supplementary

$$6x - 40 + 2x + 20 = 180$$

$$x = 25$$

Angles: 110 and 70

11.4 How Do We Classify Two-Dimensional Figures?

A two-dimensional figure is flat and has a length and a width. It does not have depth.

We are going to focus on polygons.

Polygons are closed-plane figures (start and end from the same end point), made from segments that do not cross.

Two-dimensional polygons can be classified by the number of sides.

Number of sides	Name	Example
3	Triangle	
4	Quadrilateral	
5	Pentagon	
6	Hexagon	
7	Septagon or heptagon	
8	Octagon	
9	Nonagon	
10	Decagon	

Of course, we can go on and on, but generally you only need to know the names for the common polygons. All others we usually call "20-sided polygon," etc.

Triangles

Triangles can be classified by their angles.

Acute Triangle − three acute angles

Right Triangle − one right angle

Obtuse Triangle — one obtuse angle

Triangles can also be classified by their sides.
Scalene Triangle — no congruent sides

Isosceles Triangle — at least two congruent sides

Equilateral Triangle — all congruent sides

You can classify by both angles and sides.

A triangle with a right angle and two congruent sides is called an isosceles right triangle.

Quadrilaterals

Quadrilaterals can be classified by their sides and angles as well.
Trapezoid — a quadrilateral with one pair of opposite sides parallel

Parallelogram — a quadrilateral with both pair of opposite sides congruent and parallel

Rectangle — a parallelogram with right angles

Rhombus – a parallelogram with all sides congruent

Square – a rectangle and a rhombus with all right angles and all sides congruent

We can classify the polygons by the number of sides. If they are triangles or quadrilaterals, we can use their sides and angles to give a more specific classification.

Examples:

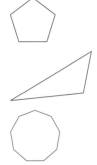

Five sides, therefore pentagon.

Three sides, therefore a triangle. Since one of the angles is obtuse, we can further classify by saying it's an "obtuse triangle."

Nine sides, therefore nonagon.

Four sides, therefore quadrilateral. Since both opposite sides are congruent and parallel, we can further classify by saying "parallelogram."

Circles

Circles are not polygons. Although a circle is a closed-plane figure, it is not made from segments. A **circle** is the set of all points in a plane having the same distance from a fixed point. The fixed point is called the **center** of the circle. The distance between the center of the circle and any point on the circle is called the **radius**. The distance from any point on the circle through the center to another point on the circle is called the **diameter**. The distance around the circle is called the **circumference**.

EXAMPLE 11.4

Classify the polygons by the number of sides and if they are triangles or quadrilaterals, use their sides and angles to give a more specific classification.

1)

2)

3)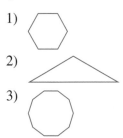

4)

5)

6)

7)

SOLUTIONS

1) hexagon
2) acute triangle
3) decagon
4) rhombus
5) trapezoid
6) scalene triangle
7) square

11.5 What Is Area and Perimeter and How Do We Calculate Them?

The area of a shape is the number that tells you how many square units are needed to cover the shape. Area can be found in different units, such as square feet, square meters, or square inches.

Example:

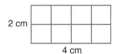

This rectangle has an area of 8 square cm.

This can be illustrated by using a square grid of centimeters; however, there are formulas you can use to calculate area.

Shape	Formula
Square	$A = s^2$
Rectangle	$A = lw$
Parallelogram Rhombus	$A = bh$
Triangle	$A = \frac{1}{2}bh$
Trapezoid	$A = \frac{1}{2}(b_1 + b_2)h$

You need to be careful to make sure you are using the correct dimensions. The height must be perpendicular to the base. You can find the area of many polygons by knowing these formulas and constructing or deconstructing the polygon into these shapes.

Examples:

Find the area of a square with a side of 9 in.

$A = s^2$
$A = (9)^2$
$A = 81 \text{ in}^2$

Find the area of a parallelogram with a height of 9.6 feet and a base of 16.2 feet.

$A = bh$
$A = (9.6)(16.2)$
$A = 155.52 \text{ feet}^2$

For example, the area of the trapezoid below can be found by using the trapezoid formula or by deconstructing the trapezoid into a rectangle and a triangle. The results will be the same. Area is preserved.

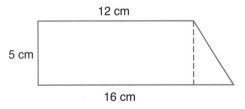

Trapezoid	Rectangle plus triangle
$\frac{1}{2}(b_1 + b_2)h$	$bh + \frac{1}{2}bh$
$\frac{1}{2}(12 + 16)5$	$12(5) + \frac{1}{2}(4)(5)$
70 cm^2	$60 + 10$
	70 cm^2

Perimeter is the distance around a polygon. You need to find the sum of the sides.

Example:

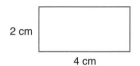

2 cm

4 cm

This rectangle has a perimeter of 12 cm.
Remember, in a rectangle, opposite sides are congruent.

Examples:

Find the perimeter of a square with a side of 9 in.
Remember, all sides are congruent.
$4(9) = 36$ 36 inches

Find the perimeter of a triangle with sides of 9.6 m, 16.2 m, and 13 m.
$9.6 + 16.2 + 13 = 38.8$ m

Find the perimeter of the trapezoid.

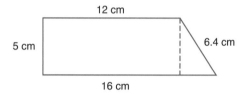

12 cm

5 cm

6.4 cm

16 cm

$5 + 12 + 16 + 6.4$
39.4 cm

CIRCLES

We can find the area of a circle using the formula: $A = \pi r^2$

For example, given a circle with a radius of 7 inches, find the area.

$A = \pi r^2$
$A = \pi(7)^2$
$A = 49\pi$ square inches

You can leave your answer in terms of π, or you can estimate your answer by substituting an approximate value for π.

$A = 49(3.14159)$
$A = 153.93791$ or 154 square inches

Another example: given a circle with a diameter of 12 m, find the area.
First, you need to find the radius. One radius is $\frac{1}{2}$ of the diameter.

$\frac{1}{2}(12) = 6$

$A = \pi r^2$

$A = \pi(6)^2$

$A = 36\pi$ square meters

You can leave your answer in terms of π or you can estimate your answer by substituting an approximate value for π.

$A = 36(3.14159)$

$A = 113.09724$ or 113 square meters

Circumference is to a circle as perimeter is to a polygon. To find the distance around a circle you will need a formula; it is not as simple as adding the sides, because there are no segments.

$C = \pi d$

For example, given a circle with a radius of 7 inches, find the circumference.
First, you need to find the diameter. One radius is $\frac{1}{2}$ of the diameter; therefore, if we double the radius, we will have the diameter.

$2(7) = 14$

$C = \pi d$

$C = \pi(14)$

$C = 14\pi$ inches

You can leave your answer in terms of π, or you can estimate your answer by substituting an approximate value for π.

$C = 14(3.14159)$

$C = 43.98226$ or 44 inches

Another example: given a circle with a diameter of 12 m, find the area.

$C = \pi d$

$C = \pi(12)$

$C = 12\pi$ meters

You can leave your answer in terms of π or you can estimate your answer by substituting an approximate value for π.

$$C = 12(3.14159)$$
$$C = 37.69908 \text{ or } 38 \text{ meters}$$

EXAMPLE 11.5

Find the area and perimeter of each polygon.

1)
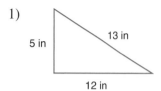

2)
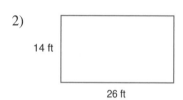

3)
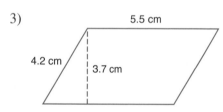

4)
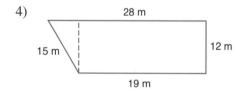

5) Find the area and circumference of the given circles.

   a) $r = 9$ ft   b) $d = 7.4$ km

SOLUTIONS

1) Perimeter: 30 in   Area: 30 in^2
2) Perimeter: 80 ft   Area: 364 ft^2
3) Perimeter: 19.4 cm   Area: 20.35 cm^2
4) Perimeter: 74 m   Area: 282 m^2
5) a) Circumference $= 18\pi$ ft   Area $= 81\pi$ ft^2
   b) Circumference $= 7.4\pi$ km   Area $= 13.69\pi$ km^2

11.6 How Do We Classify Three-Dimensional Figures?

A three-dimensional shape has length, width, and height. It is not flat. Before you can classify three-dimensional shapes, you need to know some terms.

Face a flat surface; a polygon

Edge the line where two faces meet

Vertex the point where edges intersect

Prism a solid with two congruent parallel bases and faces that are parallelograms.

The top and bottom faces (known as the *bases*) are congruent polygons, and all other faces (known as the *lateral faces*) are rectangles.

Examples cube   rectangular prism

6 faces
8 vertices
12 edges

Pyramid a solid with a base that is a polygon; the edges of the base are joined to a point outside the base. All lateral faces are triangles.

Examples triangular pyramid   square pyramid

4 faces   5 faces
4 vertices   5 vertices
6 edges   8 edges

Pyramids and prisms are named by the type of base they have.
The base is the "first name" and then pyramid or prism is the "last name."
A prism with a base of a rectangle is called a rectangular prism.

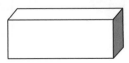

A pyramid with a base of a pentagon is called a pentagonal pyramid.

Some shapes do not have vertices, edges, or faces. These shapes have specific names.

Cylinder   a solid with two circular bases that are congruent and parallel.

Only one face; a rectangle

Cone   a solid with one circular base; the points along the circle are joined at a point outside the circle.

Sphere   a solid where every point on the surface is the same distance from the center; a 3-dimensional object shaped like a ball.

Nets of three-dimensional shapes
The net is what a 3D shape looks like if it was opened out flat or unfolded. Think about constructing a paper box. The net can then be folded up to make the shape. Here are some examples.

Net of a cube

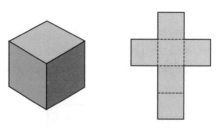

Net of a rectangular prism

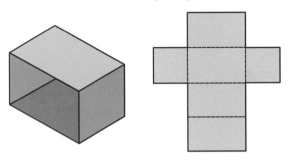

Net of a square pyramid

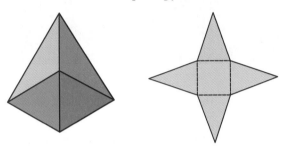

Net of a triangular pyramid

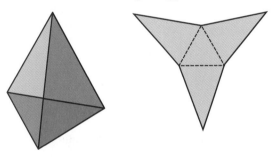

EXAMPLE 11.6 Name each of the solids.

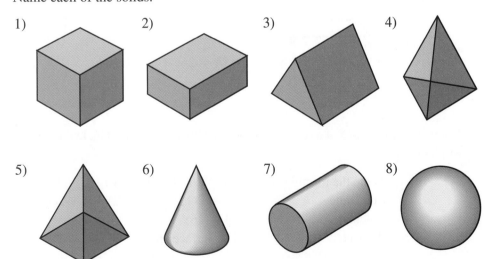

SOLUTIONS

1) cube
2) rectangular prism
3) triangular prism
4) triangular pyramid
5) square pyramid
6) cone
7) cylinder
8) sphere

11.7 What Is Surface Area and Volume and How Do We Calculate Them?

Surface Area   the area of each of the surfaces

It is helpful to think of the net of the shape when trying to picture surface area.

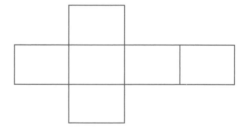

For example, if you were to paint all the faces of a rectangular prism. The surface area would be the paint.

Rectangular prism

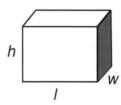

$$SA = 2lw + 2lh + 2wh$$

What is the surface area of a rectangular prism with a height of 10 in, length of 12 in, and width of 7 in?

Write the formula	$SA = 2lw + 2lh + 2wh$
Substitute	$SA = 2(12)(7) + 2(12)(10) + 2(7)(10)$
Calculate	$SA = 168 + 240 + 140$
Answer	The surface area is 548 in².

Cylinder

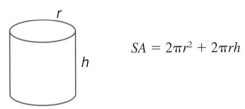

$$SA = 2\pi r^2 + 2\pi rh$$

What is the surface area of a cylinder with a height of 14 cm and radius of 5 cm?

Write the formula	$SA = 2\pi r^2 + 2\pi rh$
Substitute	$SA = 2\pi(5)^2 + 2\pi(5)(14)$
Calculate	$SA = 190\pi$
Answer	The surface area is 190π cm^2.

You can leave your answer in terms of π or you can estimate your answer by substituting an approximate value for π.

$$SA = 190(3.14159)$$
$$V = 596.9021 \text{ or } 597 \text{ cm}^2$$

Volume   is the amount of space a solid takes up. It is measured in cubic units.

If you are buying a container for storage, the label would tell you how many cubic units it holds, meaning how much stuff you can put inside the container.

Rectangular prism

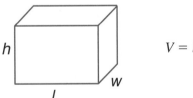

$$V = lwh$$

What is the volume of a rectangular prism with a height of 10 in, length of 12 in, and width of 7 in?

Write the formula	$V = lwh$
Substitute	$V = (12)(7)(10)$
Calculate	$V = 840$
Answer	The volume is 840 in^3.

Cylinder

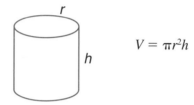

$$V = \pi r^2 h$$

What is the volume of a cylinder with a height of 14 cm and radius of 5 cm?

Write the formula $V = \pi r^2 h$

Substitute $V = \pi(5)^2(14)$

Calculate $V = 350\pi$

Answer The volume is 350π cm³.

You can leave your answer in terms of π or you can estimate your answer by substituting an approximate value for π.

$$V = 350(3.14159)$$
$$V = 1099.5565 \text{ or } 1{,}100 \text{ cm}^3$$

Cone

$$V = \tfrac{1}{3}\pi r^2 h$$
$$h = 9 \text{ in}$$
$$r = 5 \text{ in}$$

Write the formula $V = \tfrac{1}{3}\pi r^2 h$

Substitute $V = \tfrac{1}{3}\pi(5)^2(9)$

Calculate $V = 75\pi$ in³

Answer The volume is 75π in³.

You can leave your answer in terms of π or you can estimate your answer by substituting an approximate value for π.

$$V = 75(3.14159)$$
$$V = 235.619449 \text{ or } 236 \text{ in}^3$$

Sphere

$$V = \tfrac{4}{3}\pi r^3$$
$$r = 12 \text{ m}$$

Write the formula $\qquad$ $V = \frac{4}{3}\pi r^3$

Substitute $\qquad$ $V = \frac{4}{3}\pi(12)^3$

Calculate $\qquad$ $V = 2304\pi$ in^3

Answer $\qquad$ The volume is $2,304\pi$ in^3.

You can leave your answer in terms of π or you can estimate your answer by substituting an approximate value for π.

$$V = 2,304(3.14159)$$
$$V = 7,238.22336 \text{ or } 7,238 \text{ m}^3$$

 EXAMPLE 11.7 Find the volume and surface area.

1) prism $\qquad$ $l = 12$ m
   $\qquad\qquad$ $w = 4$ m
   $\qquad\qquad$ $h = 3$ m
2) cylinder $\qquad$ $r = 5$ in
   $\qquad\qquad$ $h = 13$ in

Find the volume.

3) cone $\qquad$ $r = 8$ cm
   $\qquad\qquad$ $h = 17.5$ cm
4) sphere $\qquad$ $r = 14$ ft

SOLUTIONS

1) 144 m^3 and 192 m^2
2) 325π in^3 or approx. 1,021 in^3 and 180π or approx. 565 in^2
3) $373\frac{1}{3}\pi$ cm^3 or approx. 1,173 cm^3
4) $3,658\frac{2}{3}\pi$ ft^3 or approx. 11,494 ft^3

Chapter Review

1) Find the complement of a 60 degree angle.

2) Find the supplement of a 115 degree angle.

3) Two supplementary angles are in the ratio of 2:3. Find the angles.

4) Find the missing angles.

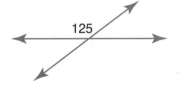
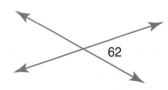

5) Line *l* is parallel to line *m*. Find the missing angles.

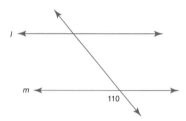

6) Line *l* is parallel to line *m*. Solve for *x*.

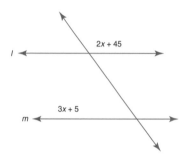

7) Find the area and perimeter.

 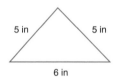

Height = 4 in

8) Find the area and circumference (in terms of π)

Radius = $3\frac{1}{2}$ feet

9) Find the volume and surface area to the nearest tenth.

diameter = 8 in   length = 15.6 cm
height = 7 in   width = 2.3 cm
                          height = 3.4 cm

10) Natalie needs to get a bigger fish tank. Her current tank does not hold enough
    fish. Her current tank holds 10 liters. The new tank she wants is 40 cm by 15 cm
    by 25 cm. Will the new tank be big enough? Justify your answer.
    (1,000 cubic cm = 1 liter)

Functions and Graphing

WHAT YOU WILL LEARN

- To produce a table, equation, and graph of a linear function.
- To identify the slope of a line as positive or negative from a graph.
- To determine the slope of a linear function, given an equation not necessarily in slope-intercept form.
- Interpret the equation $y = mx + b$ as defining a linear function, whose graph is a straight line.
- Compare properties of two functions, each represented in a different way (algebraically, graphically, numerically in tables, or by verbal descriptions).
- A function is a rule that assigns to each input exactly one output.
- About domain and range, and which is allowed to have numbers repeat.
- To use the vertical line test to determine a function.
- To determine if functions are linear or non-linear.
- To identify linear functions from their equation, table, and graph.

SECTIONS IN THIS CHAPTER
• What Is Coordinate Geometry?
• How Do We Graph a Line from a Table of Values?
• What Is Slope?
• How Do We Graph a Line Using Slope?
• What Is a Function?
• What Is the Rule of Four?
• How Can We Tell If It Is Linear or Non-linear?

Axis A horizontal or vertical line used in the Cartesian coordinate system to locate a point on the coordinate graph.

Axes The horizontal and vertical lines dividing a coordinate plane into four quadrants.

Cartesian (coordinate) plane The plane formed by a horizontal axis and a vertical axis, often labeled the x-axis and y-axis respectively. The point of intersection of the axes is called the origin and has coordinates (0,0).

Coordinates An ordered pair of numbers that identifies a point on a coordinate plane, written as (x,y).

Origin The point on the coordinate plane where the x- and y-axes intersect; has coordinates (0,0).

Quadrant One of four sections of a coordinate grid separated by horizontal and vertical axes; they are numbered I, II, III, and IV, counterclockwise from the upper right.

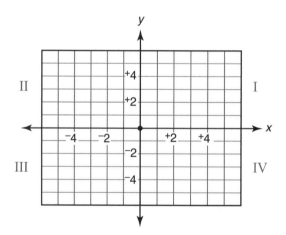

x-axis The horizontal axis; the line whose equation is $y = 0$.

x-intercept The point where a graph of an equation crosses the x-axis.

y-axis The vertical axis; the line whose equation is $x = 0$.

y-intercept The point where a graph of an equation crosses the y-axis.

12.1 What Is Coordinate Geometry?

The **x-axis** runs horizontally, which is left and right.

The **y-axis** is perpendicular to the x-axis and runs vertically, which is up and down. The axes divide the coordinate plane into four equal regions called quadrants. They are numbered with Roman numerals by starting at (+,+) the top-right quadrant and moving in a counterclockwise direction. The points on the plane are found by using ordered pairs of (x,y). The x always comes before the y, just like in the alphabet.

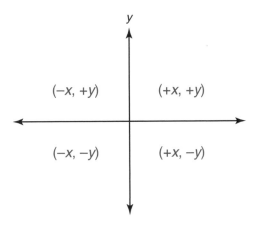

Remember, the intersection of two lines is called a point. You can use lines or your finger if it helps. Think, "Find the elevator," then go up or down.

The point (2, 3) is located by moving from the origin 2 to the right and 3 up. An ordered pair gives direction and distance. The direction is from the sign and the distance is the number.

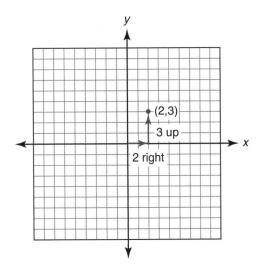

1) Graph and label each of the following points on the coordinate plane. Name the quadrant.

    A (2, 3) is located in Quadrant ____.

    B (−4, −5) is located in Quadrant ____.

    C (−5, 3) is located in Quadrant ____.

    D (4, −3) is located in Quadrant ____.

    E (0, 0) is located at the _____.

2) Name the ordered pair of each of the given points.

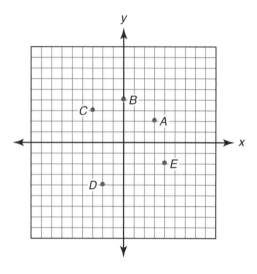

A _____

B _____

C _____

D _____

E _____

SOLUTIONS

1) A (2, 3) is located in Quadrant I.

    B (−4, −5) is located in Quadrant III.

    C (−5, 3) is located in Quadrant II.

    D (4, −3) is located in Quadrant IV.

    E (0, 0) is located at the origin.

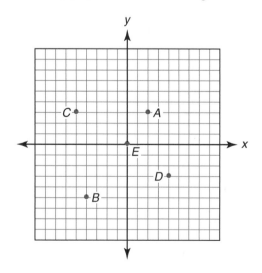

2) A $(3, 2)$
    B $(0, 4)$
    C $(-3, 3)$
    D $(-2, -4)$
    E $(4, -2)$

12.2 How Do We Graph a Line from a Table of Values?

Table of values   An organized list of values from a function/relation.

You can graph functions from their equations or from a table of values. We are going to work with the table of values.

Look at the equation $y = 2x + 1$ in a table of values. The x-values are the independent variables. You choose them. Choose numbers that you are comfortable using and that will fit nicely on your paper. You might love the number 10, but do you really want to graph that large a number?

Take each x-value and substitute to find the y-value. The y-values are the dependent variables. The y-value depends on the x-value you substitute.

x	y
1	3
2	5
3	7
4	9

$y = 2(1) + 1 = 3$
$y = 2(2) + 1 = 5$
$y = 2(3) + 1 = 7$
$y = 2(4) + 1 = 9$

Now that you have your ordered pairs, graph them. Connect to make a long line, and label with the equation.

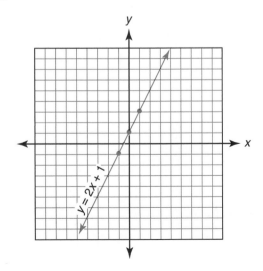

Look at the equation $y = 3x - 5$ in a table of values.

Choose your x-values. It is always good to pick numbers in a pattern. Put a pattern in, get a pattern out!

Take each x-value and substitute to find the y-value.

x	y
0	-5
1	-2
2	1
3	4

$$y = 3(0) - 5 = -5$$
$$y = 3(1) - 5 = -2$$
$$y = 3(2) - 5 = 1$$
$$y = 3(3) - 5 = 4$$

Now that you have your ordered pairs, graph them. Connect to make a long line, and label with the equation.

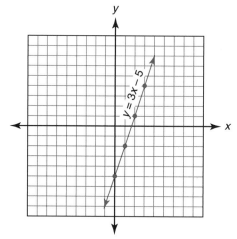

EXAMPLE 12.2 Graph the function by creating a table of values.

1) $y = \frac{1}{2}x + 3$

2) $y = 4x + 6$

3) $y = 3x - 7$

4) $y = -x + 6$

SOLUTIONS

1)

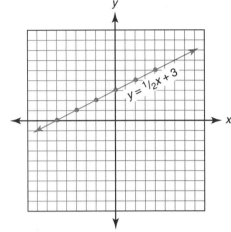

x	y
0	3
2	4
4	5
6	6

2)

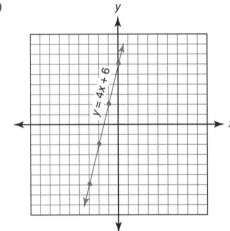

x	y
−2	−2
−1	2
0	6
1	10

3)

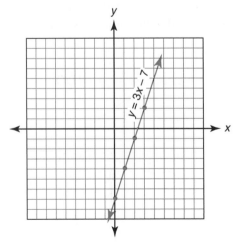

x	y
0	−7
1	−4
2	−1
3	2

4)

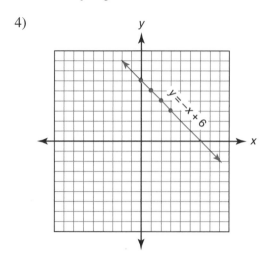

x	y
0	6
1	5
2	4
3	3

12.3 What Is Slope?

slope   The measure of the steepness of a line; the ratio of vertical change to horizontal change.

Example:

If point P is (x_1, y_1) and point Q is (x_2, y_2) the slope of $\overline{PQ}$ is $\dfrac{\Delta y}{\Delta x} = \dfrac{y_2 - y_1}{x_2 - x_1}$.

The slope is steepness, which is like incline. Think about walking up a hill: some are harder than others—it is usually all about the steepness. With slope, the sign of the slope will give you the direction and the number will determine the steepness. The higher the number, the steeper the line.

Positive slope goes UP

Negative slope goes DOWN

Zero slope is horizontal

Undefined slope is vertical

Slope calculations are found by finding the ratio of change—the difference in the y-coordinates compared to the difference in the x-coordinates.

Look at the picture.
The *y*-value increases by 4.
The *x*-value increases by 3.

The slope is $\frac{4}{3}$.
The change in *y* compared to the change in *x*.

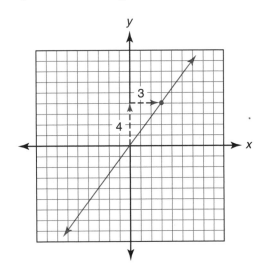

Using subtraction

$(1, 3)$ and $(-2, 4)$

To find the change in *y*: subtract 3 from 4 to get 1.
To find the change in *x*: subtract 1 from -2 to get -3.

The slope is $-\frac{1}{3}$.

$(-2, 5)$ and $(4, 1)$

To find the change in *y*: subtract 1 from 5 to get 4.
To find the change in *x*: subtract 4 from -2 to get -6.

The slope is $-\frac{4}{6}$ or $-\frac{2}{3}$.

 EXAMPLE 12.3   Identify the slope as positive, negative, zero, or undefined.

1)

2)
3)

4)

You can use either graphing and counting or subtraction to find the slope.

5) $(2, -5)$ and $(-6, -5)$
6) $(3, 6)$ and $(1, -2)$
7) $(-1, 2)$ and $(5, 5)$
8) $(-1, -1)$ and $(-3, 2)$

SOLUTIONS

1) positive
2) zero
3) negative
4) undefined
5) zero
6) $\frac{4}{1}$ or 4
7) $\frac{3}{6}$ or $\frac{1}{2}$
8) $-\frac{3}{2}$

12.4 How Do We Graph a Line Using Slope?

If you have one point and the slope, you can find an infinite number of other points on the same line. The slope tells us the rate of change, so we know how the x and y values increase or decrease together.

For example, if you start at point $(1, 2)$ and move using a slope of 3 you can complete the table.

Increases by 1

x	y
1	2
2	5
3	8
4	11

Increases by 3

Now you can use the table of values to graph the line.
You can also use the equation.

Slope-intercept form   The equation of a straight line in the form $y = mx + b$, where m is the slope and b is the y-coordinate of the point where the line intercepts the y-axis.

$$y = 2x - 3$$
Slope y-intercept

This requires the equation to be in $y =$ form. If so, it is that simple.

If an equation is linear (first-degree x) then you can use slope-intercept form to find the slope. Remember, slope is a ratio, so you are going to want to write the slope as a fraction.

- Write the equation in $y =$ form ($y = mx + b$)
- Find the coefficient of x.
  - This number is the slope.
  - If there is no x, there are zero x.
- Find the constant.
  - This number is the y-intercept.
  - If there is no constant, it is zero.

Knowing the y-intercept gives you a point $(0, b)$. You can choose to continue to make a chart starting with the y-intercept as your first number, or go straight to the graph. The y-intercept is where you begin and the slope is how you move. Remember: m for move and b for begin.

$$y = 2x - 3$$

The slope is 2. Think 2 up / 1 right.

The y-intercept is -3. Begin at $(0, -3)$ and move 2 up, 1 right. Usually you will want to make at least three points before you draw the line. Label your line with the equation.

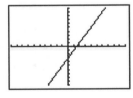

$$y = 3x + 1$$

The slope is 3. Think 3 up / 1 right.

The y-intercept is 1. Begin at (0,1) and move 3 up, 1 right. Usually you will want to make at least three points before you draw the line. Label your line with the equation.

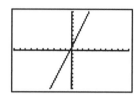

$$y = \tfrac{1}{2}x + 4$$

The slope is $\tfrac{1}{2}$. Think 1 up / 2 right.

The y-intercept is 4. Begin at (0, 4) and move 1 up, 2 right. Usually you will want to make at least three points before you draw the line. Label your line with the equation.

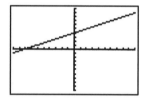

EXAMPLE 12.4

Name the slope and y-intercept and graph the line.

1) $y = 3x - 5$

2) $y = \tfrac{1}{2}x + 3$

3) $y = -2x + 4$

4) $y = \tfrac{2}{3}x - 1$

5) $2y = 4x + 8$ (get the y alone first)

SOLUTIONS

1) Slope:3 y-intercept: -5

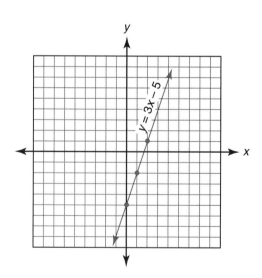

2) Slope: $\frac{1}{2}$ y-intercept: 3

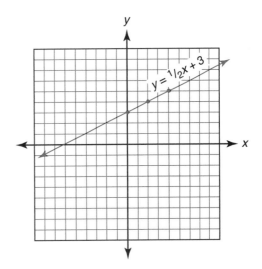

3) Slope: -2 y-intercept: 4

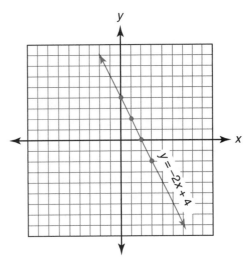

4) Slope: $\frac{2}{3}$ y-intercept: -1

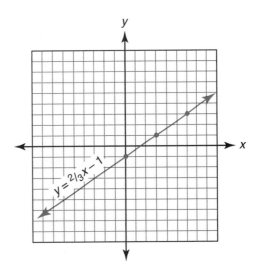

5) Slope: 2   *y*-intercept: 4

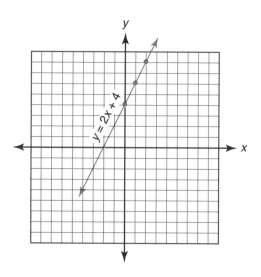

12.5 What Is a Function?

Function   A mathematical relationship between two variables, an independent variable and a dependent variable, where every value of the independent variable corresponds to exactly one value of the dependent variable.

Think of a function as a machine. The machine takes in a number, performs operations, and puts out another number. There are certain conditions that are set on the machine in order for it to be a function machine. You can call it input and output or *x*-values and *y*-values, but there is also new vocabulary.

Domain of a function   The set of input values of a function.

Range of a function   The set of output values of a function.

Domain input	Range output
x	$f(x)$
x	y

All the same ideas, just different names.

In a function, each element of the domain is paired with exactly one element in the range.

$$\{(1, 2), (2, 4), (3, 5)\}$$

x	y
1	5
2	7
3	9
4	11

You can represent a relation as a table, graph, rule (equation), or in words.

functions

x	y
1	2
2	3
3	4

x	y
1	4
2	5
3	6

x	y
1	1
2	4
3	9

x	y
1	4
2	4
3	4

NOT functions

x	y
1	2
1	3
2	4

x	y
1	4
2	5
1	6

Think: Does it function?

It is okay for the *y*-value to repeat.
For example, $y = 2$ is a function.

It is not okay for the *x*-values to repeat.
For example, $x = 2$ is not a function.

All functions must pass the **vertical line test**.

The vertical line test is done by drawing a vertical line through the function. If the vertical line has only one point of intersection, then the graph is a function.

Vertical line tests

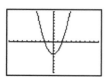

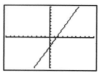

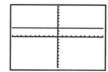

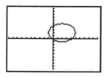

Yes, function   Yes, function   Yes, function   No, not a function

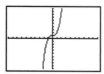

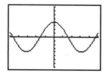

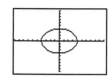

Yes, function   Yes, function   No, not a function   No, not a function

Another difference with functions is the way you write them. The domain is still written using x; however, the y is written using $f(x)$.

Function notation   A notation in which a function is named with a letter and the input is shown in parentheses after the function name.

$$f(x) = x^2 + 1 \text{ represents the function } y = x^2 + 1$$

EXAMPLE 12.5

For exercises 1–4, answer the following:
What is the domain?
What is the range?
Is it a function?

1) {(1, 2), (2, 4), (3, 6)}

2)

x	y
1	5
2	7
3	9
4	11

3) {(0, 2), (2, 4), (0, 5)}

4)

x	y
1	5
2	3
3	3
4	5

5) Which of the following pass the vertical line test?

a)

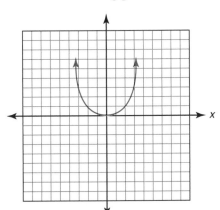

b)

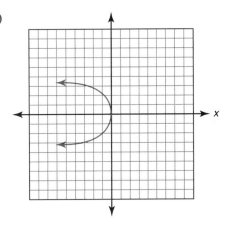

c)

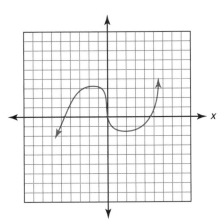

d)

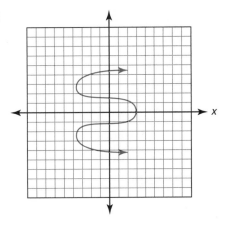

e)

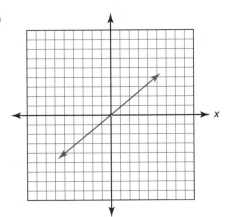

f)

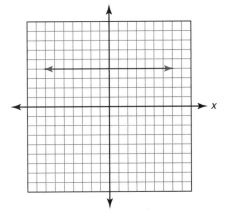

SOLUTIONS

1) Domain: 1,2,3
   Range: 2,4,6
   Function: yes
2) Domain: 1,2,3,4
   Range: 5,7,9,11
   Function: yes

3) Domain: 0,2
   Range: 2,4,5
   Function: No
4) Domain: 1,2,3,4
   Range: 3,5
   Function: yes
5) a) yes
   b) no
   c) yes
   d) no
   e) yes
   f) yes

12.6 What Is the Rule of Four?

Basically, the rule of four states that any mathematical relationship can be expressed four ways.

- Algebraically
- Arithmetically
- Graphically
- Verbally

Here is an example:

Algebraically

$y = 3x - 5$

Verbally

The rule of y is three times the independent variable x decreased by five.

Arithmetically

x	y
1	−2
2	1
3	4
4	7
5	10

Graphically

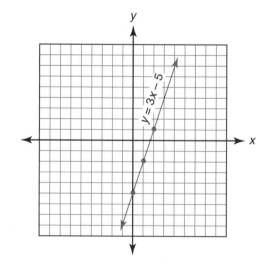

Here is another example.

Algebraically

$y = \frac{1}{2}x + 2$

Verbally

The rule of y is half the independent variable x increased by two.

Arithmetically

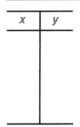

Graphically

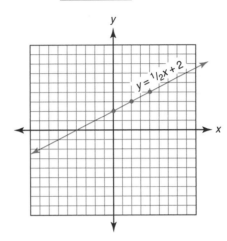

You can look at functions from their equations **or** from a table of values **or** from their graphs **or** their verbal descriptions.

EXAMPLE 12.6 Fill in the missing parts of the rule of four.

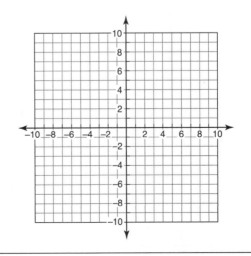

1)

Algebraically	Verbally
	The rule of y is twice the independent variable x, decreased by seven.

Arithmetically

Graphically

2)

Algebraically	Verbally
$2y = 10x - 8$	

Arithmetically	Graphically

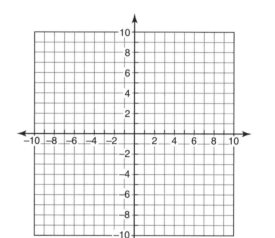

3)

Algebraically	Verbally

Arithmetically	Graphically

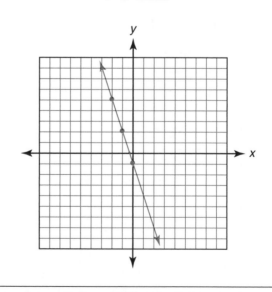

SOLUTIONS

1) $y = 2x - 7$

The rule of y is twice the independent variable x decreased by seven.

x	y
1	−5
2	−3
3	−1
4	1
5	3

2) $y = -2x + 5$

The rule of y is negative two times the independent variable x decreased by five.

x	y
0	5
1	3
2	1
3	−1
4	−3

3) $y = -3x - 1$

The rule of y is the product of negative three and the independent variable x decreased by one.

x	y
0	−1
1	−4
2	−7
3	−10
4	−13

12.7 How Can We Tell If It Is Linear or Non-linear?

First, the easiest way is to look at the graph. Linear equations are straight; they have a constant rate of change or slope.

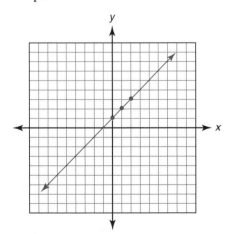

Second, look at the variables of the equation. A linear equation is 1st degree. The exponents are 1. If the variable has an exponent of any number other than 1, then it cannot be a line.

Third, we are going to work with the **table of values**.

For a 1st degree equation, the first difference will be constant. The equation will be a straight line. We call these linear equations. This first difference is the slope.

x	y
1	3
2	5
3	7
4	9

For a 2nd degree equation, the second difference will be constant. The equation will be a parabola. A parabola is a curve, not a line; it is non-linear.

x	y
1	1
2	4
3	9
4	16
5	25

There are many other examples of non-linear equations that you will need to know later in math. For now, we are focusing on only 1st degree equations (linear) and 2nd degree equations (non-linear).

We will use the "difference" to determine if the equation is linear.

Later on in math, we will use the "difference" to help us write the rules.

x	y
−2	2
0	0
2	−2
4	−4

$\Delta x = 2$ $\Delta y = -2$ linear

x	y
1	−4
2	−2
3	0
4	2

$\Delta x = 1$ $\Delta y = 2$ linear

x	y
1	2
2	−2
3	−4
4	−8

$\Delta x = 1$ Δy is not constant   non-linear

x	y
0	1
1	2
2	5
3	10

$\Delta x = 1$ Δy is not constant   non-linear

However, there is a pattern, so let's try the second difference. The second difference is 2, so is a constant. It is still non-linear; however, it is second degree, which means it will be a parabola.

Remember:

You can look at their graphs. If you are uncertain, try a ruler!

You can also tell from the equation. Look at the exponents.

Tell whether linear or non-linear.

1) $y = 2x - 6$

2) $y = \frac{1}{2}x + 4$

3) $y = x^2 - 9$

4)

x	y
−1	4
0	3
1	2
2	1

5)

x	y
−1	2
0	5
1	10
2	17

SOLUTIONS

1) linear

2) linear

3) non-linear

4) linear

5) non-linear

Chapter Review

Graph the line using the table of values.

1) $y = 2x - 5$

2) $y = \frac{1}{2}x + 3$

Calculate the slope.

3) $(0, -8)$ and $(-5, 2)$

4) $(3, 4)$ and $(3, 9)$

5) $(-4, 6)$ and $(2, -1)$

6) Complete the table.

Equation	Slope	y-intercept
$y = 4x - 18$		
$y = -2x$		
	2	-7
	$\frac{1}{3}$	0
	8	-3

7) State the domain, range, and whether it is a function.
    $\{(0,3), (1,5), (2,7)\}$

8) Tell whether the equation is linear or non-linear.
    a) $y = 3x - 7$
    b) $y = x^2 - 6$
    c) $y = 2$
    d) $y = x^3$

Transformational Geometry

WHAT YOU WILL LEARN

- The properties of dilations, rotations, reflections, and translations.
- How to describe the effect of dilations, translations, rotations, and reflections on two-dimensional figures using coordinates.

SECTIONS IN THIS CHAPTER

- What Is Transformational Geometry?
- What Is a Reflection?
- What Is a Translation?
- What Is Rotation?
- What Is a Dilation?
- What Is Symmetry?

13.1 What Is Transformational Geometry?

The word transform means "to change." A transformation can be a change in size or location. People often think of skateboarders or figure skaters as they do flips or turns.

Transformation   The result of a change made to an object.

The three basic transformations are:
Reflection (flip)

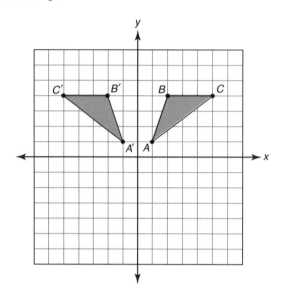

Rotation (turn)

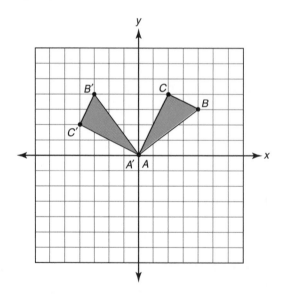

Translation (slide)

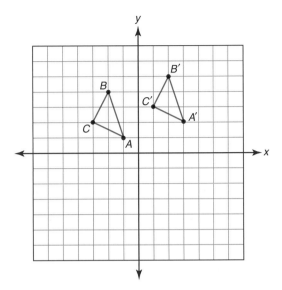

All of these transformations change the location of an object, but each in a different way.

The fourth transformation is dilation. A dilation is different because it changes the location and the size.

Dilation (shrink or stretch)

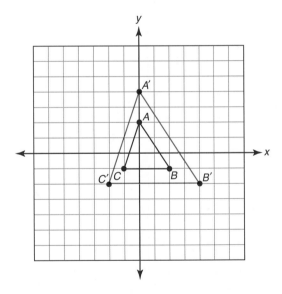

The vocabulary is simple because it is pretty familiar to most of us. The original figure is the pre-image. The transformed figure is called the image.

Image The figure created when another figure, called the pre-image, undergoes a transformation.

Pre-image In transformational geometry, the figure before a transformation is applied.

Preserved In transformational geometry, a property that is kept or maintained.

EXAMPLE
13.1

Match the transformation to the illustration.

1. Reflection
2. Rotation
3. Translation
4. Dilation

A.

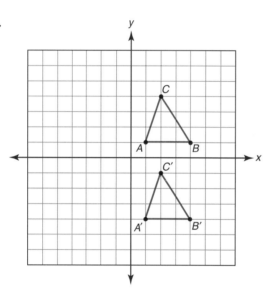

C.

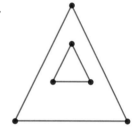

B.

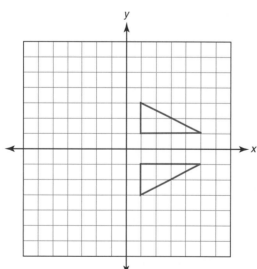

D.

SOLUTIONS
1. B
2. D
3. A
4. C

13.2 What Is a Reflection?

When reflecting a figure, every point has a corresponding point on the other side of the line of symmetry—like when you look in the mirror or a very still body of water, you see your reflection.

Reflection   In transformational geometry, the figure formed by flipping a geometric figure (pre-image) over a line (the "line of symmetry") to obtain a mirror image; informally known as a "flip." (The line is the perpendicular bisector of each line segment connecting a point and its image. The mirror image is the same distance from the line of symmetry as the original.)

Line of symmetry   A line that divides a figure into two congruent halves that are mirror images of each other; a simple test to determine if a figure has line symmetry is to fold the figure along the supposed line of symmetry and see if the two halves of the figure coincide.

Example:
Reflect △ABC over the x-axis.

A $(1, 2) \rightarrow A'\ (1, -2)$
B $(2, 4) \rightarrow B'\ (2, -4)$
C $(3, 1) \rightarrow C'\ (3, -1)$

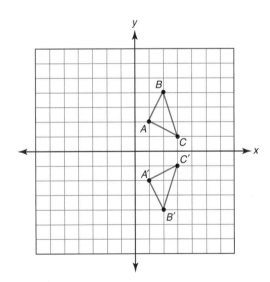

When you reflect over the x-axis, the sign of the y-value will change.

Example:

Reflect $\triangle ABC$ over the y-axis.

$A \quad (1, 2) \rightarrow A' \, (-1, 2)$

$B \quad (2, 4) \rightarrow B' \, (-2, 4)$

$C \quad (3, 1) \rightarrow C' \, (-3, 1)$

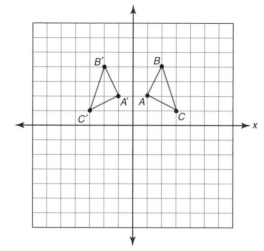

When you reflect over the y-axis, the sign of the x-value will change.

You can do a reflection over any line; the idea is that the pre-image and the image are an equal distance from the line of reflection or the line of symmetry. The image and the pre-image are congruent; we say that "size is preserved."

EXAMPLE 13.2

1) Graph $\triangle ABC$ and then draw the image of ABC after a reflection over the x-axis. Label the image $A'B'C'$.

$A \quad (1, -3)$

$B \quad (3, -5)$

$C \quad (2, -8)$

2) Graph $\triangle ABC$ and then draw the image of ABC after a reflection over the y-axis. Label the image $A'B'C'$.

$A \quad (-2, 1)$

$B \quad (-5, 5)$

$C \quad (-6, 1)$

SOLUTIONS

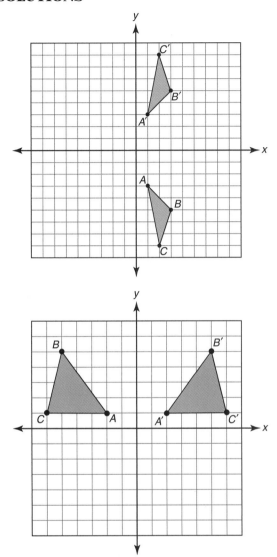

13.3 What Is a Translation?

A translation is when every point is moved in the same direction the same distance. If you have ever seen synchronized swimming or the "electric slide" dance, you get the idea. Everything has to move at the same time for the same distance—otherwise they will bump into each other.

Translation A transformation where every point moves the same distance in one or two directions within the plane; informally known as a "slide."

Example:

Translate $\triangle ABC$ 6 units to the left.

   A $(1, 2) \rightarrow A'\,(-5, 2)$
   B $(2, 4) \rightarrow B'\,(-4, 4)$
   C $(3, 1) \rightarrow C'\,(-3, 1)$

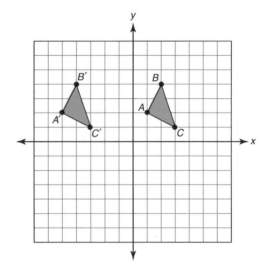

When you translate left, you subtract from the x-value.

Example:

Translate $\triangle ABC$ 6 units to the right.

   A $(1, 2) \rightarrow A'\,(7, 2)$
   B $(2, 4) \rightarrow B'\,(8, 4)$
   C $(3, 1) \rightarrow C'\,(9, 1)$

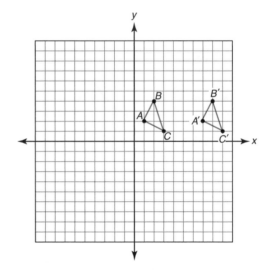

When you translate right, you add to the x-value.

Example:

Translate $\triangle ABC$ 6 units down.

A $(1, 2) \rightarrow A'$ $(1, -4)$
B $(2, 4) \rightarrow B'$ $(2, -2)$
C $(3, 1) \rightarrow C'$ $(3, -5)$

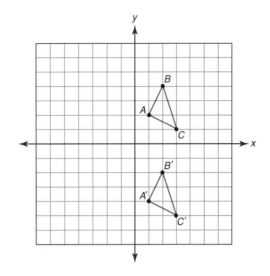

When you translate down, you subtract from the y-value.

Example:

Translate $\triangle ABC$ 6 units up.

A $(1, 2) \rightarrow A'$ $(1, 8)$
B $(2, 4) \rightarrow B'$ $(2, 10)$
C $(3, 1) \rightarrow C'$ $(3, 7)$

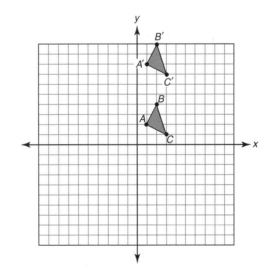

When you translate up, you add to the y-value.

When you create a translation, the idea is that the pre-image is moved to create the image. You may be given one direction or two. Generally, you would move left or right before moving up or down, similar to the way you graph a point. The image and the pre-image are congruent; we say that "size is preserved."

EXAMPLE 13.3

1) Graph $\triangle ABC$ and then draw the image of ABC after a translation of 4 left and 2 up. Label the image $A'B'C'$.

A $(1, -3)$
B $(5, -7)$
C $(2, -9)$

2) Graph $\triangle ABC$ and then draw the image of ABC after a translation of 5 right and 3 up. Label the image $A'B'C'$.

   A $(-1, -3)$
   B $(-5, -7)$
   C $(-2, -9)$

SOLUTIONS

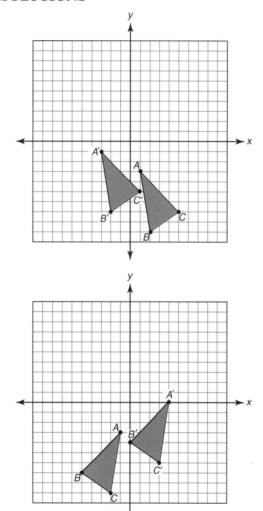

13.4 What Is Rotation?

The earth rotates on its axis. A ballerina rotates on her toes while doing a pirouette. When you rotate, you turn. The rotation needs to have a focal point or center of rotation.

Rotation   A transformation or movement that results when a geometric figure is rotated about a fixed point; informally known as a "turn."

Rotational symmetry   A figure has rotational symmetry when it can be rotated around a central point or point of rotation less than 360° and still be identical to the original figure.

For a rotation, you need to be given a direction and a number of degrees. Usually the degrees are in multiples of 90, but it can be any amount of degree. If no direction is indicated, then you would move counterclockwise—the same way the quadrants are labeled. The center of rotation is the origin.

Example:

Rotate $\triangle ABC$ 90° counterclockwise.

$A \quad (1, 2) \rightarrow A'\ (-2, 1)$
$B \quad (2, 4) \rightarrow B'\ (-4, 2)$
$C \quad (3, 1) \rightarrow C'\ (-1, 3)$

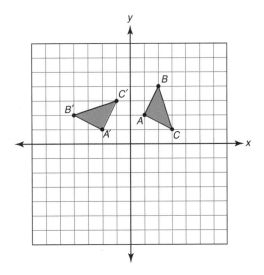

When you rotate 90°, the x-value and y-value switch. The rule is to also make the new x-value negative; however, most prefer to get the positive and negative signs by knowing the quadrant. Since you are now in quadrant II, the x-values are negative.

Note: Rotating 90° counterclockwise is the same as rotating 270° clockwise.

Example:

Rotate $\triangle ABC$ 180° counterclockwise.

A $(1, 2) \rightarrow A'\,(-1, -2)$
B $(2, 4) \rightarrow B'\,(-2, -4)$
C $(3, 1) \rightarrow C'\,(-3, -1)$

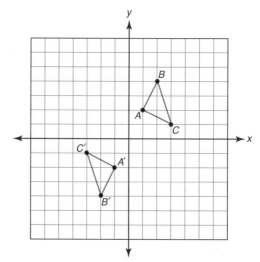

When you rotate 180°, the *x*-value and *y*-value do not switch. You are basically rotating by 90° twice, so it would be switch then switch back. Since you are now in quadrant III, the *x*-values and *y*-values are negative.

Note: Rotating 180° counterclockwise is the same as rotating 180° clockwise.

Example:

Rotate $\triangle ABC$ 270° counterclockwise.

A $(1, 2) \rightarrow A'\,(2, -1)$
B $(2, 4) \rightarrow B'\,(4, -2)$
C $(3, 1) \rightarrow C'\,(1, -3)$

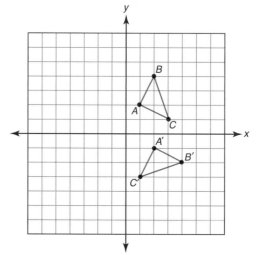

When you rotate 270 degrees, the *x*-value and *y*-value switch. Since you are now in quadrant IV, the *y*-values are negative.

Note: Rotating 270° counterclockwise is the same as rotating 90° clockwise.

When you create a rotation, the idea is that the pre-image is turned to create the image. The image and the pre-image are congruent; we say that "size is preserved."

1) Graph $\triangle ABC$ and then draw the image of ABC after a rotation of 180° counterclockwise. Label the image $A'B'C'$.

   A $(-1, -3)$
   B $(-3, -5)$
   C $(-4, -2)$

2) Graph $\triangle ABC$ and then draw the image of ABC after a rotation of 90° counterclockwise. Label the image $A'B'C'$.

   A $(-1, 2)$
   B $(-1, 5)$
   C $(-5, 1)$

SOLUTIONS

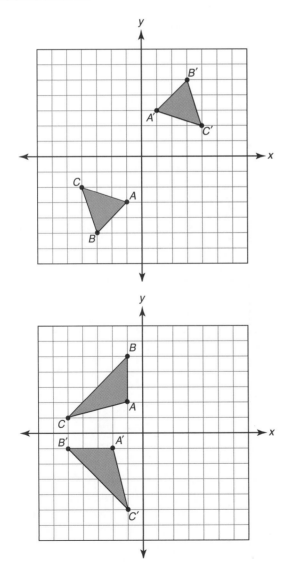

13.5 What Is a Dilation?

Have you ever gone to an eye doctor and he places drops in your eyes? The drops stop your eyes from dilating when light is seen. Normally, your eyes dilate to control the amount of light that enters the eye. Pupils become smaller when the light is bright, and larger in the dark. Eye doctors dilate eyes to check for signs of disease. This does not change the shape of your pupils, just the size. The same holds true for a dilation in mathematics.

Dilation A transformation that stretches or shrinks a function or graph both horizontally and vertically by the same scale factor.

Example:

Dilate $\triangle ABC$ by a scale factor of 3.

$A\ \ (1, 2) \rightarrow A'\ (3, 6)$

$B\ \ (2, 4) \rightarrow B'\ (6, 12)$

$C\ \ (3, 1) \rightarrow C'\ (9, 3)$

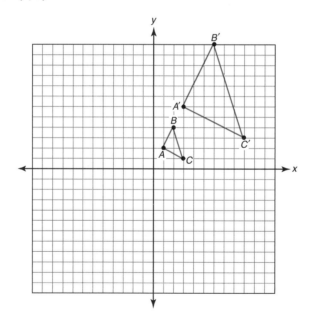

When you dilate, you multiply both the x-value and the y-value by the scale factor. When the scale factor is greater than one, the figure is enlarged.

Example:

Dilate $\triangle ABC$ by a scale factor of $\frac{1}{3}$.

$A \quad (-3, 3) \to A' \, (-1, 1)$

$B \quad (0, 6) \to B' \, (0, 2)$

$C \quad (3, 3) \to C' \, (1, 1)$

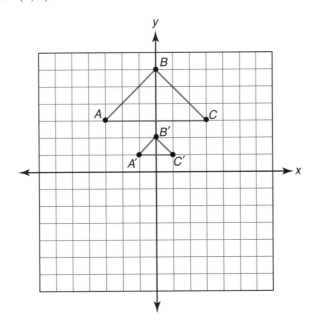

When you dilate, you multiply both the *x*-value and the *y*-value by the scale factor. When the scale factor is greater than one, the figure is enlarged. When you create a dilation, the idea is that the pre-image is made bigger or smaller, depending on the scale factor. The image and the pre-image are not congruent, but the shapes are similar or in proportion.

1) Graph $\triangle ABC$ and then draw the image of ABC after a dilation of a scale factor of $\frac{1}{4}$. Label the image $A'B'C'$.

$A \quad (-8, 0)$

$B \quad (0, 8)$

$C \quad (4, 4)$

2) Graph $\triangle ABC$ and then draw the image of ABC after a dilation of a scale factor of 2. Label the image $A'B'C'$.

$A \quad (-1, -3)$

$B \quad (0, 2)$

$C \quad (3, 1)$

SOLUTIONS

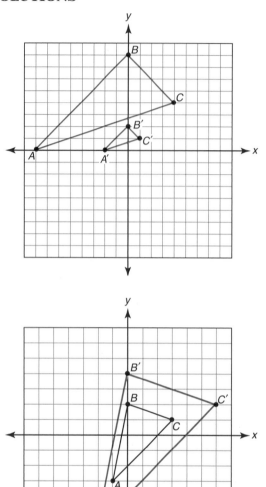

13.6 What Is Symmetry?

When you look at a Valentine's Day heart, it appears to have two halves. The shape is said to have symmetry. Also, if you look at a pinwheel, it appears to have the same shape repeated; it has symmetry as well.

Symmetry   The property of having the same size and shape across a dividing line or around a point.

There are two types of symmetry. The simplest symmetry is reflection symmetry (line or mirror symmetry); it is easy to see because a line divides the shape in half. One half is the reflection of the other.

Harder to find is rotation or point symmetry. The image has been rotated about a point. When creating a design, the rotation may occur several times, for example, rotating 90° an image three times about the center of rotation to create an image appearing all the way around a center.

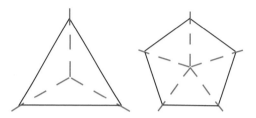

Let's look at the alphabet for symmetry. The red lines demonstrate horizontal line symmetry. The black lines demonstrate vertical line symmetry. Imagine folding a letter along a line; the halves will match up. The circled letters have point symmetry. Turn them upside down to see.

A	B	C	D
E	F	G	⊞
⊕	J	K	L
M	N	⊕	P
Q	R	S	†
ψ	V	W	✳
Y	Z		

Chapter Review

1) Graph $\triangle ABC$ and then draw the image of ABC after a reflection over the y-axis. Label the image $A'B'C'$.

A $(-1, -3)$

B $(-3, -5)$

C $(-2, -8)$

2) Graph $\triangle ABC$ and then draw the image of ABC after a reflection over the x-axis. Label the image $A'B'C'$.

A $(3, -6)$

B $(1, 2)$

C $(5, 1)$

3) Graph $\triangle ABC$ and then draw the image of ABC after a translation of 3 left and 5 up. Label the image $A'B'C'$.

A $(-2, -3)$

B $(0, -7)$

C $(2, -1)$

4) Graph $\triangle ABC$ and then draw the image of ABC after a dilation of scale factor $\frac{1}{2}$. Label the image $A'B'C'$.

A $(-4, -4)$

B $(0, 8)$

C $(-6, 2)$

5) Graph $\triangle ABC$ and then draw the image of ABC after a dilation of scale factor 3. Label the image $A'B'C'$.

A $(-2, -2)$

B $(0, 3)$

C $(2, -2)$

6) Graph $\triangle ABC$ and then draw the image of ABC after a rotation of $270°$ counter-clockwise. Label the image $A'B'C'$.

A $(0, 3)$

B $(2, 5)$

C $(5, 1)$

7) Which of the following letters have vertical line symmetry?

A   B   C   D   E   F   H   I

8) Which of the following letters have horizontal line symmetry?

A   B   C   D   E   F   H   I

System of Linear Equations

WHAT YOU WILL LEARN

- Analyze and solve pairs of simultaneous linear equations
- Understand the solutions to a system of two linear equations in two variables correspond to points of intersection of their graphs

SECTIONS IN THIS CHAPTER

- What Are the Possibilities When Solving Simultaneous Linear Equations?
- How Do We Solve a System Graphically?
- How Do We Solve a System by Addition?
- How Do We Solve a System by Substitution?
- How Do We Solve Word Problems Leading to Two Linear Equations?

14.1 What Are the Possibilities When Solving Simultaneous Linear Equations?

A system of linear equations consists of two or more linear equations. A solution to the system is any ordered pair that is a solution in both or all equations.

Our focus will be on linear equations, therefore all graphs will be lines. Ask yourself: when two lines cross, how many times do they cross? What if they don't cross? What if they lay on top of one another?

Think about vertical angles for a moment. Vertical angles are formed when two lines intersect.

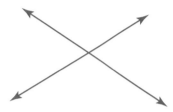

When two lines intersect at one point, the system is called a consistent system. The two lines have different slopes. There is exactly one solution.

Well, now think about parallel lines. Parallel lines do not intersect.

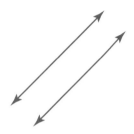

When two lines are parallel, the system is called inconsistent. The two lines have the same slope. There is no solution.

Now think about drawing a line and then tracing over the same line. They are the exact same line.

When two lines are the same, the system is called dependent. The two lines have the same slope and the same y-intercept. There are an infinite number of solutions.

We can determine which type of system we have by looking at the graphs or by looking at the equations in $y =$ form.

For example, let's look at the equations below.

$$y = 3x - 5$$
$$y = \tfrac{1}{3}x - 2$$
$$y = 3x + 7$$
$$y = 3x - 5$$

The first equation, $y = 3x - 5$, will intersect the equation $y = \tfrac{1}{3}x - 2$ because their slopes are different. This system is consistent. The first equation $y = 3x - 5$ will be parallel to the equation $y = 3x - 7$. They have the same slope but different y-intercepts. This system would be inconsistent. The first equation $y = 3x - 5$ is the same as $y = 3x - 5$. They have the same slope and the same y-intercept. This system is dependent.

Sometimes, you will need to put the equations in $y =$ form or slope-intercept form in order to tell the type of system.

For example,

$$4x - 2y = 8 \quad \text{and} \quad x + 3y = 12$$

First, get the equations in $y =$ form.

$$y = 2x - 4 \quad \text{and} \quad y = -\tfrac{1}{3}x + 4$$

By looking at the slopes, you should be able to determine the type of system. The slopes are different; this system is consistent and the graph will be a pair of intersecting lines.

 EXAMPLE 14.1

Tell which type of system and the number of solutions.

1) $y = 2x - 8$
   $y = -2x - 8$
2) $2y = 4x + 12$
   $y = 4x - 5$
3) $2x + y = 5$
   $3y = -6x + 9$
4) $4x - y = -9$
   $-4x + y = 9$

SOLUTIONS

1) Consistent; one solution $(0, -8)$
2) Consistent; one solution $(5.5, 17)$
3) Inconsistent; no solution
4) Dependent; infinite solutions

14.2 How Do We Solve a System Graphically?

To solve a system of equations graphically, graph each of the equations in the system. The solution(s), if any, will be the ordered pair(s) of the point(s) of intersection of all the graphs.

Example:

Solve the following system of equations graphically:

$$y = \tfrac{1}{6}x + 3$$
$$y = \tfrac{2}{3}x$$

1) Graph the first equation
2) Graph the second equation

3) Find the intersection of the lines
4) Name the coordinates (the solution set of the system of equations)
5) Check in both equations

$(6, 4)$

Check

$y = \frac{1}{6}x + 3$ $y = \frac{2}{3}x$

$(4) = \frac{1}{6}(6) + 3$ $(4) = \frac{2}{3}(6)$

$4 = 1 + 3$ $4 = 4$

$4 = 4$

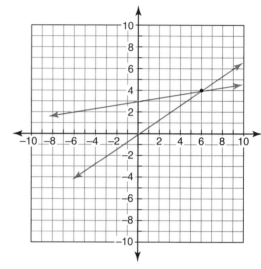

Solve the following system of equations graphically.

$$y = 2x - 4$$
$$3y = 6x + 12$$

When changed into $y =$ form, the second equation is

$$y = 2x + 4.$$

The lines will be parallel.

There is no solution.

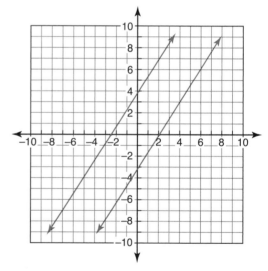

EXAMPLE 14.2

1) Solve the following system of equations graphically:
   $y = 5x - 3$
   $y = x + 5$
2) Solve the following system of equations graphically and check.
   $y = -2x + 1$
   $y = x + 7$
3) Solve the following system of equations graphically and check.
   $3y = -2x + 9$
   $2x + 3y = 9$

SOLUTIONS

1)

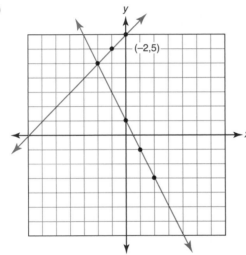

Consistent system; solution is (2, 7)

2)

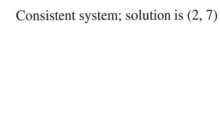

Consistent system; solution is (−2, 5)

3)

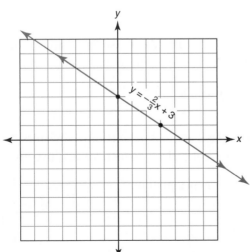

Dependent system; infinite solutions.

14.3 How Do We Solve a System by Addition?

Since graphing is not always practical, you may have a large or non-integer solution set. We will look now at other ways to solve the system. In order to solve by addition, we need to add the two equations together to make one equation. The key to making one equation is getting the equation to contain only one variable. In order for a variable to "drop out," we need additive inverses. We need for one of the variables to have a sum of zero.

$$\text{For example,} \quad y + x = 7$$
$$y - x = 11$$

If we were to combine the two equations, the x and $-x$ would have a sum of zero; we would be creating a new equation with only one variable.

Make sure to line up the equal signs when you add.

$$y + x = 7$$
$$+ \quad y - x = 11$$
$$\overline{\quad\quad 2y = 18}$$
$$y = 9$$

Now that we have a y value, we can go back to either of the original equations and find the value of x by substitution.

$$y + x = 7$$
$$(9) + x = 7$$
$$x = -2$$

The solution to the equation is $(-2, 9)$.

Check the solution by making sure the values make both equations true.

$$y + x = 7 \qquad\qquad y - x = 11$$
$$(9) + (-2) = 7 \qquad\qquad (9) - (-2) = 11$$
$$7 = 7 \qquad\qquad 11 = 11$$

The equations are not always given to you in a ready-to-use form. Sometimes, we have to use least common multiples to make the equations contain additive inverses.

$$\text{For example,} \quad 2x - 6y = 2$$
$$-4x + 8y = -8$$

Let's look at the x-variables. We need the 2 and -4 to be additive inverses. If we multiply the first equation by 2, we will have 4 and -4, which are additive inverses.

$$2(2x - 6y = 2) \qquad\qquad \text{new equation:} \qquad 4x - 12y = 4$$
$$-4x + 8y = -8 \qquad\qquad\qquad\qquad\qquad + \quad -4x + 8y = -8$$
$$\overline{\qquad\qquad\qquad -4y = -4}$$
$$y = 1$$

Now that we have a y value, we can go back to either of the original equations and find the value of x by substitution.

$$2x - 6y = 2$$
$$2x - 6(1) = 2$$
$$2x - 6 = 2$$
$$2x = 8$$
$$x = 4$$

The solution to the equation is (4, 1).

Check the solution by making sure the values make both equations true.

$$
\begin{array}{ll}
2x - 6y = 2 & -4x + 8y = -8 \\
2(4) - 6(1) = 2 & -4(4) + 8(1) = -8 \\
8 - 6 = 2 & -16 + 8 = -8 \\
2 = 2 & -8 = -8
\end{array}
$$

Another example,
$$3x + 2y = 6$$
$$x + 3y = -5$$

Let's look at the y-variables. We need the 2 and 3 to be additive inverses. If we multiply the first equation by a 3 and the second equation by a -2, we will have 6 and -6 which are additive inverses.

$$3(3x + 2y = 6)$$
$$-2(x + 3y = -5)$$

new equations:

$$
\begin{array}{r}
9x + 6y = 18 \\
+ \quad -2x - 6y = 10 \\
\hline
7x = 28 \\
x = 4
\end{array}
$$

Now that we have an x-value, we can go back to either of the original equations and find the value of y by substitution.

$$x + 3y = -5$$
$$4 + 3y = -5$$
$$3y = -9$$
$$y = -3$$

The solution to the equation is (4, -3).

Check the solution by making sure the values make both equations true.

$$
\begin{array}{ll}
3x + 2y = 6 & x + 3y = -5 \\
3(4) + 2(-3) = 6 & (4) + 3(-3) = -5 \\
12 + -6 = 6 & 4 + -9 = -5 \\
6 = 6 & -5 = -5
\end{array}
$$

EXAMPLE 14.3

Find the solution set.

1) $2x - y = 2$
   $x + y = 7$
2) $5x + y = 13$
   $4x - 3y = 18$
3) $7x + 2y = 5$
   $y = 16 - 2x$
4) $x + y = 12$
   $x - y = 4$

SOLUTIONS

1) $(3, 4)$
2) $(3, -2)$
3) $(-9, 34)$
4) $(8, 4)$

14.4 How Do We Solve a System by Substitution?

When solving a system by substitution, the result is an equation formed by using both equations. You need to solve one of the equations for a single variable. In other words, solve for either x or y. Once you have the x or y alone, you take the resulting expression and use it to replace the variable in the other equation.

For example, we have two equations:

$$4x + 3y = 27$$
$$y = 2x - 1$$

We are going to replace the y in the first equation with the expression $2x - 1$, creating a new equation with only one variable.

$$4x + 3(2x - 1) = 27$$

Now we will solve the new equation.

$4x + 6x - 3 = 27$	Use the distributive property
$10x - 3 = 27$	Combine like terms
$10x = 30$	
$x = 3$	

Now that we have an x-value, we can go back to either of the original equations and find the value of y by substitution.

$$y = 2x - 1$$
$$y = 2(3) - 1$$
$$y = 6 - 1$$
$$y = 5$$

The solution to the equation is $(3, 5)$.

Check the solution by making sure the values make both equations true.

$$
\begin{array}{c@{\qquad}c}
4x + 3y = 27 & y = 2x - 1 \\
4(3) + 3(5) = 27 & (5) = 2(3) - 1 \\
12 + 15 = 27 & 5 = 6 - 1 \\
27 = 27 & 5 = 5
\end{array}
$$

Another example:

This time, we will have to do a little work to isolate the variable. Let's solve for x with the second equation.

$$3x - 4y = 26$$
$$x + 2y = 2 \qquad \text{New equation:} \quad x = -2y + 2$$

We are going to replace the x in the first equation with the expression $-2y + 2$, creating a new equation with only one variable.

$$3(-2y + 2) - 4y = 26$$

Now we will solve the new equation.

$$
\begin{array}{ll}
-6y + 6 - 4y = 26 & \text{Use the distributive property} \\
-10y + 6 = 26 & \text{Combine like terms} \\
-10y = 20 & \\
y = -2 &
\end{array}
$$

Now that we have a y-value, we can go back to either of the original equations and find the value of x by substitution.

$$x + 2y = 2$$
$$x + 2(-2) = 2$$
$$x + -4 = 2$$
$$x = 6$$

The solution to the equation is $(6, -2)$.

Check the solution by making sure the values make both equations true.

$$
\begin{array}{c@{\qquad}c}
3x - 4y = 26 & x + 2y = 2 \\
3(6) - 4(-2) = 26 & (6) + 2(-2) = 2 \\
18 + 8 = 26 & 6 - 4 = 2 \\
26 = 26 & 2 = 2
\end{array}
$$

EXAMPLE 14.4   Find the solution set.

1) $y = 2x$
   $x + y = 21$
2) $y = x + 1$
   $x + y = 9$
3) $7x - 3y = 23$
   $x + 2y = 13$
4) $y - x = -2$
   $3x - y = 16$

SOLUTIONS

1) $(7, 14)$
2) $(4, 5)$
3) $(5, 4)$
4) $(7, 5)$

14.5 How Do We Solve Word Problems Leading to Two Linear Equations?

You can solve many real-world problems using a system of linear equations. You need to use the story to write two equations, then choose any of the methods (graphically or algebraically) to solve.

Example:

Johnny bought a large popcorn and 3 chocolate chip cookies for $5.00. Gab bought a large popcorn and 5 chocolate cookies for $6.00. How much is one popcorn and one cookie?

Set up some equations.

Let x = popcorn
Let y = cookie

Johnny's equation:	$x + 3y = 5$
Gab's equation:	$x + 5y = 6$

Choose a method to solve. For this system, we will demonstrate substitution.

First, solve for x. $\qquad\qquad x = -3y + 5$
Second, substitute $\qquad (-3y + 5) + 5y = 6$
Solve $\qquad\qquad\qquad\qquad 2y + 5 = 6$
$\qquad\qquad\qquad\qquad\qquad 2y = 1$
$\qquad\qquad\qquad\qquad\qquad x = .50$

Substitute back into one of the original equations.

$$x + 3(.50) = 5$$
$$x + 1.50 = 5$$
$$x = 3.50$$

The popcorn is $3.50 and the cookies are $.50 each.

Another example, when Beck cashed a check for $170, the bank teller gave him 12 bills, some $20 bills and the rest $10 bills. How many bills of each denomination did he receive?

Set up some equations.

Let x = the number of ten-dollar bills
Let y = the number of twenty-dollar bills

Using the total number of bills: $\qquad\qquad x + y = 12$
Using the denominations and total value: $\qquad 10x + 20y = 170$
Choose a method to solve. For this system, we will demonstrate addition.

First, we will use the additive inverses of -10 and 10 therefore, we will multiply the first equation by -10.

$$x + y = 12 \qquad\qquad \text{New equation:} \quad -10x - 10y = -120$$
$$10x + 20y = 170 \qquad\qquad\qquad\qquad +\quad \underline{10x + 20y = 170}$$

Add the two equations together. $\qquad\qquad\qquad\qquad\qquad\qquad 10y = 50$
Solve $\qquad\qquad\qquad\qquad\qquad\qquad\qquad\qquad\qquad\qquad\qquad y = 5$
Substitute back into one of the original equations.

$$x + y = 12$$
$$x + 5 = 12$$
$$x = 7$$

There are 7 ten-dollar bills and 5 twenty-dollar bills.

Another example: Phone company A charges $35 per month plus $.25 per minute roaming charges. Phone company B charges $45 per month and $.05 per minute roaming charges. When will these plans cost the same?

Set up some equations.

Let x = the number of minutes roaming
Let y = the total bill per month

$$\text{Company A:} \qquad y = .25x + 35$$
$$\text{Company B:} \qquad y = .05x + 45$$

Choose a method to solve. For this system, we will demonstrate graphing.

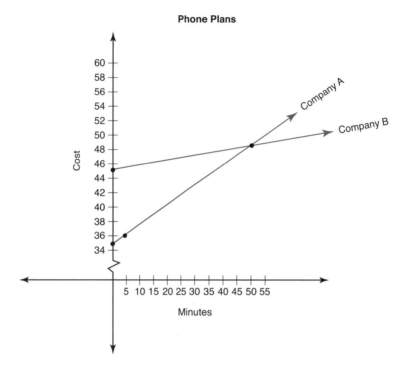

They will cost the same at 50 minutes. They will both cost $47.50.

EXAMPLE 14.5

Solve each of the word problems using any method.

1) Joshua paid $26.50 for 10 gallons of gas and 2 quarts of oil. Anthony paid $23.58 for 8 gallons of gas and 3 quarts of oil. Find the cost of 1 gallon of gas and 1 quart of oil.

2) Katie rented two movies and three video games for a total cost of $24.30. Ryan rented three movies and one game for a total cost of $18.25. Find the price of one game and one movie.

3) Angela bought 3 belts and 3 scarves for $33. Valerie bought 2 belts and 1 scarf for $19. Find the price of one belt and one scarf.

4) Joey has 6 whiffle balls and 4 bats that weigh 3.5 kilograms. Julia has 20 whiffle balls and 12 bats that weigh 11 kilograms. They want to ship them individually. How much does each bat and ball weigh?

SOLUTIONS

1) Gasoline cost $2.31 per gallon and oil cost $1.70 per quart

2) A movie costs $4.35 and a game costs $5.20

3) Belts cost $6.00 each and scarves cost $5.00 each

4) Whiffle balls weigh .25 kilograms and bats weigh .5 kilograms.

Chapter Review

Name the type of solution. If the system is consistent, find the solution set.

1) $5x - 2y = 20$
   $2x + 3y = 27$

2) $3y - 6x = 12$
   $5y = 10x + 20$

3) $3x + 7y = -2$
   $2x + 3y = -3$

4) $x + y = 6$
   $9 - x = y$

5) Solve by the substitution method.
   $4x - y = 10$
   $2x + 3y = 12$

6) Solve by the addition method.
   $5x + 8y = 1$
   $3x + 4y = -1$

7) Solve by graphing.
   $y = 2x - 7$
   $x + y = 5$

8) A florist is selling roses and carnations. A bouquet with 5 roses and 3 carnations costs $19.50. A bouquet with 2 roses and 10 carnations costs $21.00. What is the cost per flower?

The Pythagorean Theorem

WHAT YOU WILL LEARN

- How to explain a proof of the Pythagorean Theorem and its converse.
- How to apply the Pythagorean Theorem to determine unknown side lengths in right triangles in real-world and mathematical problems in two and three dimensions.
- How to apply the Pythagorean Theorem to find the distance between two points in a coordinate system.

SECTIONS IN THIS CHAPTER

- What Is the Pythagorean Theorem?
- What Is the Converse of the Pythagorean Theorem?
- How Can We Use the Pythagorean Theorem?
- How Can We Use the Pythagorean Theorem on the Coordinate Plane?

15.1 What Is the Pythagorean Theorem?

A long time ago, a man named Pythagoras discovered a very interesting property of right triangles. He was able to prove that it worked for all right triangles, and since it was proved to work for all, it became a theorem. First, it has to be a right triangle, which means the triangle contains a right angle.

Hypotenuse The side of a right triangle opposite the right angle; the longest side of a right triangle.

Example:

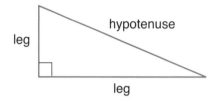

Leg(s) of a right triangle One of the two sides that form the right angle of a right triangle; the sides that are not the hypotenuse (see hypotenuse).

Pythagorean Theorem The mathematical relationship stating that in any right triangle, the sum of the squares of the two legs is equal to the square of the hypotenuse; if a and b are the lengths of the legs and c is the length of the hypotenuse, then $a^2 + b^2 = c^2$.

This means that if you take the sides of triangles and make a square on each of the three sides, then the biggest square will have the exact same area as the other two squares put together.

There are a few triangles whose sides are Pythagorean triples. A triple is when the sides are all integers. Here are some examples:

3, 4, 5	because $3^2 + 4^2 = 5^2$
5, 12, 13	because $5^2 + 12^2 = 13^2$
7, 24, 25	because $7^2 + 24^2 = 25^2$

Also, any multiple of a set of triples will work. It would be a similar triangle that has been dilated.

15.2 What Is the Converse of the Pythagorean Theorem?

A converse is when you switch the front and the back of a statement.

The Pythagorean Theorem states:

> If a triangle is a right triangle, then the sum of the squares of the legs is equal to the square of the hypotenuse.

If we switch it, it becomes:

> If the sum of the squares of legs is equal to the square of the hypotenuse, the triangle is a right triangle.

This is commonly used to verify if a triangle is a right triangle.

Example:

Are 6, 8, 10 the sides of a right triangle?

We would use the converse to test it.

If $6^2 + 8^2 = 10^2$ then we can state the triangle is a right triangle.

$$6^2 + 8^2 = 10^2$$
$$36 + 64 = 100$$
$$100 = 100 \qquad \text{True}$$

It is a right triangle.

Example:

Are 11, 13, 21 the sides of a right triangle?

We would use the converse to test it.

If $11^2 + 13^2 = 21^2$ then we can state the triangle is a right triangle.

$$11^2 + 13^2 = 21^2$$
$$121 + 169 = 441$$
$$290 = 441 \qquad \text{False}$$

It is NOT a right triangle.

EXAMPLE 15.2

State whether the triangle is a right triangle.
1) 9, 12, 15
2) 6, 7, 12
3) 12, 16, 20
4) 9, 10, 14

SOLUTIONS

1) Yes
2) No
3) Yes
4) No

15.3 How Can We Use the Pythagorean Theorem?

We have seen how we can use the Pythagorean Theorem to tell whether or not a triangle is a right triangle. If we know the triangle is a right triangle, then we can use the Pythagorean Theorem to find one of the missing sides if the other two are known.

Example:

If we know the legs are 30 and 40, find the hypotenuse.

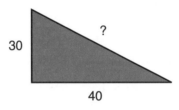

$$a^2 + b^2 = c^2$$ Write down the formula.

$$(30)^2 + (40)^2 = c^2$$ Substitute.

$$900 + 1600 = c^2$$ Evaluate.

$$2500 = c^2$$ Add.

$$\sqrt{2500} = \sqrt{c^2}$$ Take the square root of each side.

$$c = 50$$

Not all answers will be integers; only the Pythagorean triples will have integers as answers. Depending on the question, you could leave the answer as a radical or an estimate.

Example:

If we know the legs are 4 and 19, find the hypotenuse.

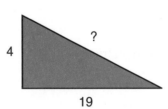

$$a^2 + b^2 = c^2$$ Write down the formula.

$$(4)^2 + (19)^2 = c^2$$ Substitute.

$$16 + 361 = c^2$$ Evaluate.

$$376 = c^2 \qquad \text{Add.}$$
$$\sqrt{376} = \sqrt{c^2} \qquad \text{Take the square root of each side.}$$
$$c = \sqrt{376} \qquad \text{If you leave the radical answer.}$$
$$c = 19.4 \qquad \text{Rounded to nearest tenth.}$$

It may be helpful to draw a diagram to visualize the problem. It will all work out if you switch the legs as the commutative property will protect the answer. You can't switch the hypotenuse though, because it would not work out.

Example:

If we know one leg is 10 and the hypotenuse is 26, find the missing leg.

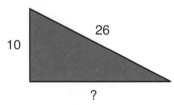

$$a^2 + b^2 = c^2 \qquad \text{Write down the formula.}$$
$$(10)^2 + b^2 = (26)^2 \qquad \text{Substitute.}$$
$$100 + b^2 = 676 \qquad \text{Evaluate.}$$
$$-100 = -100 \qquad \text{Subtract from each side.}$$
$$b^2 = 576 \qquad \text{Take the square root of each side.}$$
$$b = \sqrt{576} \qquad \text{If you leave the radical answer.}$$
$$b = 24 \qquad \text{Integer answer.}$$

Example:

A cat is stuck up a tree. We have a 20-ft. ladder which is being placed 4 feet away from the base of the tree. Will they be able to reach the cat? If we know one leg is 4 and the hypotenuse is 20, find the missing leg.

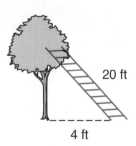

$$a^2 + b^2 = c^2 \qquad \text{Write down the formula.}$$
$$(4)^2 + b^2 = (20)^2 \qquad \text{Substitute.}$$
$$16 + b^2 = 400 \qquad \text{Evaluate.}$$
$$-16 = -16 \qquad \text{Subtract from each side.}$$
$$b^2 = 384 \qquad \text{Take the square root of each side.}$$
$$b = \sqrt{384} \qquad \text{If you leave the radical answer.}$$
$$b = 19.6 \qquad \text{Rounded to the nearest tenth.}$$

We can reach the cat as long as the cat is 19.6 feet high or less.

EXAMPLE 15.3

1) Find the missing side (round to the nearest tenth if needed).

a)
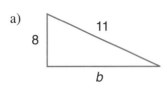

b)
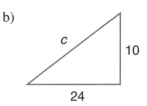

c)
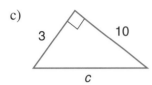

d)
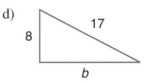

2) What is the height of the tent?

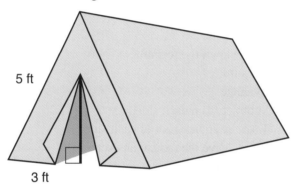

3) What is the height of the wheel chair ramp to the nearest tenth?

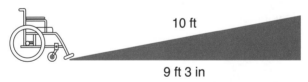

SOLUTIONS

1) a) $b = 7.5$
   b) $c = 26$
   c) $c = 10.4$
   d) $b = 15$
2) 4 feet
3) 3.8 feet

15.4 How Can We Use the Pythagorean Theorem on the Coordinate Plane?

Let's look at the graph of the three points.

A $(0, 0)$

B $(0, 3)$

C $(4, 0)$

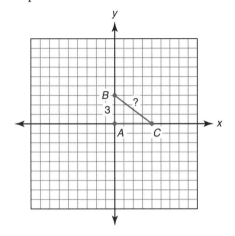

The length of AB is 3.

The length of AC is 4.

How can we find the length of BC?

If we look at the points and relate them to the vertices of a triangle, we will notice that it is a right triangle. Since it is a right triangle, we can use the Pythagorean Theorem to find side BC.

$a^2 + b^2 = c^2$	Write down the formula.
$(3)^2 + (4)^2 = c^2$	Substitute.
$9 + 16 = c^2$	Evaluate.
$25 = c^2$	Add.
$\sqrt{25} = \sqrt{c^2}$	Take the square root of each side.
$c = 5$	

The length of BC is 5.

We can use this concept to find the distance between any two points as long as we create the right triangle.

Let's look at the graph of the two points.

A $(-2, 3)$

B $(2, 2)$

Connect to make a right triangle using (2, 3) as the third point.

$AC = 4$

$BC = 1$

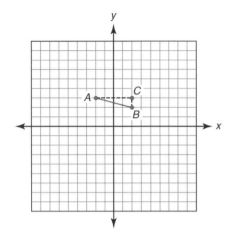

$a^2 + b^2 = c^2$	Write down the formula.
$(1)^2 + (4)^2 = c^2$	Substitute.
$1 + 16 = c^2$	Evaluate.
$17 = c^2$	Add.
$\sqrt{17} = \sqrt{c^2}$	Take the square root of each side.
$c = \sqrt{17}$	If you leave the radical answer.
$c = 4.1$	Rounded to nearest tenth.

Chapter Review

1) State whether the triangle is a right triangle.
    a) 20, 21, 29
    b) 36, 48, 60
    c) 15, 16, 21

2) Find the missing side (to the nearest tenth).

a)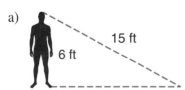

15 ft

6 ft

b)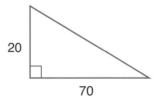

20

70

c)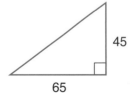

45

65

d)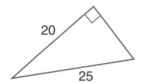

20

25

3) Write and solve an equation that can be used to find the distance in miles between the two planes.

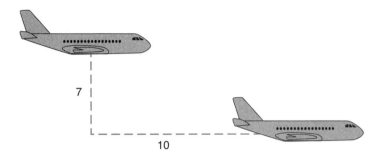

4) Carmelo is parasailing. The tow rope is 40 feet long and Carmelo is a horizontal distance of 25 feet from the back of the boat. How high is Carmelo?

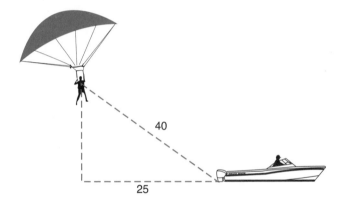

5) Graph each of the ordered pairs. Then find the distance between the two points.
   a) (3, 0) and (7, −5)
   b) (2, 0) and (5, −4)
   c) (1, 3) and (−2, 4)

Probability

WHAT YOU WILL LEARN

- How to use tree diagrams or the fundamental counting principle to count outcomes.
- How to represent sample spaces of simple and compound events.
- That the probability of a chance event is a number between 0 and 1 that expresses the likelihood of the event occurring.
- How to find probabilities of simple events.
- How to approximate the probability of a chance event by collecting data on the chance process and predicting the approximate relative frequency given the probability.
- How to find probabilities of compound events using organized lists, tables, tree diagrams, and simulation.
- That the probability of a compound event is the fraction of outcomes in the sample space for which the compound event occurs.
- How to set expectations for certain events.

SECTIONS IN THIS CHAPTER

- What Are Tree Diagrams and the Counting Principle?
- What Is Probability?
- How Do We Find the Probability of Simple Events?
- How Do We Find the Probability of Compound Events?
- What Is a Permutation?
- What Is a Combination?

16.1 What Are Tree Diagrams and the Counting Principle?

Event A set of one or more outcomes in a probability experiment.

Compound event A combination of two or more simple events.

A sample space is a set of all possible outcomes for an activity or experiment. It will be a list or a diagram.

Rolling a die, then tossing a coin.

1,	head	1,	tail
2,	head	2,	tail
3,	head	3,	tail
4,	head	4,	tail
5,	head	5,	tail
6,	head	6,	tail

When attempting to determine a sample space to list the outcomes from an experiment, it is often helpful to draw a diagram that illustrates how to arrive at the answer.

One such diagram is a tree diagram or a list—this could be a list of ordered pairs. Some people do both the diagram and the list.

Tree diagram

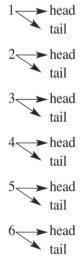

Example:

Show the sample space and tree diagram for rolling a pair of dice.

Tree Diagram:

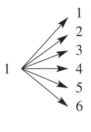

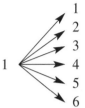

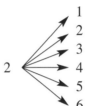

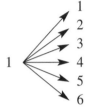

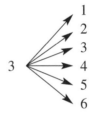

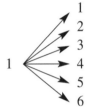

Sample Space:

1,1	2,1	3,1	4,1	5,1	6,1
1,2	2,2	3,2	4,2	5,2	6,2
1,3	2,3	3,3	4,3	5,3	6,3
1,4	2,4	3,4	4,4	5,4	6,4
1,5	2,5	3,5	4,5	5,5	6,5
1,6	2,6	3,6	4,6	5,6	6,6

Counting principle   The fundamental counting principle states that if event M can occur m ways and is followed by event N that can occur in n ways, then event M followed by N can occur in $m \times n$ ways.

Example:
If a shirt comes in 3 colors and 4 sizes, then there are 3×4 or 12 possible outcomes.

This simple multiplication process is known as the counting principle.

The counting principle works for two or more activities.

We can look at the sample space:

red, small	blue, small	green, small
red, medium	blue, medium	green, medium
red, large	blue, large	green, large
red, extra large	blue, extra large	green, extra large

Rather than list the entire sample space with all possible combinations, we may simply multiply $3 \times 4 = 12$ possible choices.

Example:
A coin is tossed five times. How many arrangements of heads and tails are possible?

2 choices (heads or tails) and 5 tosses

$2 \times 5 = 10$ possible outcomes

Example:
You have 5 skirts and 6 shirts. How many outfits can you make?

5 skirts and 6 shirts

$5 \times 6 = 30$ outfits

1) You have three types of bread, four meats, two cheeses, and four toppings. How many different kinds of sandwiches can you make?
2) You have six types of ice cream and four toppings. How many sundaes can you make?
3) Draw a tree diagram or sample space.
   a) Tossing a quarter and then a dime.
   b) Pulling 2 marbles from a bag of 2 red, 1 blue, and 1 black. One marble is drawn and then replaced.
   c) Pulling 2 marbles from a bag of 2 red, 1 blue, and 1 black. One marble is drawn and then NOT replaced.

SOLUTIONS
1) $3 \times 4 \times 2 \times 4 = 96$ sandwiches
2) $6 \times 4 = 24$ sundaes

3) a) Q(head), D(head)
        Q(head), D (tail)
        Q (tail), D (head)
        Q (tail), D (tail)

    b) red, red   red, red   blue, red   black, red
        red, red   red, red   blue, red   black, red
        red, blue   red, blue   blue, blue   black, blue
        red, black   red, black   blue, black   black, black

    c) red, red   red, red   blue, red   black, red
        red, blue   red, blue   blue, red   black, red
        red, black   red, black   blue, black   black, blue

16.2 What Is Probability?

Probability   The chance of an event occurring; the ratio of the number of favorable outcomes to the total number of possible outcomes; the probability of an event must be greater than or equal to 0 and less than or equal to 1.

In addition to helping determine the number of outcomes in a sample space, the tree diagram can be used to determine the probability of individual outcomes within the sample space.

The probability of any outcome in the sample space is the product (multiplication) of all possibilities along the path that represents that outcome on the tree diagram. These events are usually independent, but can be dependent.

The probability of an event falls between impossible and certain.

$$0 \leq \text{probability of an event} \leq 1$$

This is usually written as a ratio, but it can be written as a decimal or a percentage.

A probability of 0 means the event is impossible or that outcome will never happen.

A probability of 1 means the event is certain or that outcome will surely happen.

The closer a probability is to 1, the more likely the outcome.

The closer the probability is to 0, the less likely the outcome.

A sample space will generally help us determine possibilities.

**EXAMPLE
16.2**
Determine if the event is impossible, unlikely, likely, or certain.

1) Snow in October in New York
2) Rain in May in Virginia
3) Winning the million-dollar lottery
4) Winning the lottery if you did not play
5) Having school on a Tuesday in December
6) A unicycle having one wheel

SOLUTIONS

1) unlikely
2) likely
3) unlikely
4) impossible
5) likely
6) certain

16.3 How Do We Find the Probability of Simple Events?

Larger numbers indicate a greater likelihood. A probability near 0 indicates an unlikely event, a probability around $\frac{1}{2}$ indicates an event that is neither unlikely nor likely, and a probability near 1 indicates a likely event.

Example:

If a student is selected at random from a class, find the probability that Adriana will be selected. Adriana has 22 students in her class. She is one out of the 22 students. The probability she is chosen is $\frac{1}{22}$.

Another Example:

Matty has a game tonight. He plays soccer, basketball, and lacrosse. The probability that the game is a basketball game is $\frac{1}{3}$.

The ratio that describes the chance that a particular event will occur is called probability. The numerator of the ratio is the number of ways in which the particular outcome could occur, or the number of successes.

The denominator of the ratio is the number of ways in which all the different outcomes could occur, or the total number of outcomes.

$$\text{Probability} = \frac{\text{Number of successes}}{\text{Total possibilities}}$$

Example:

A spinner is made up of three equal regions labeled 1, 2, and 3.

The sample space is 1, 2, 3.

What is the probability of the following events?

a) $P(1) = \frac{1}{3}$

b) $P(2) = \frac{1}{3}$

c) $P(\text{not } 3) = \frac{2}{3}$

d) $P(\text{even}) = \frac{1}{3}$

e) $P(\text{odd}) = \frac{2}{3}$

f) $P(\text{a number less than } 4) = \frac{3}{3} = 1$

g) $P(\text{a number greater than } 3) = \frac{0}{3} = 0$

Example:

A letter is chosen at random from the word MISSISSIPPI.

sample space: M   IIII   PP   SSSS

What is the probability of the following events?

a) $P(\text{I}) = \frac{4}{11}$

b) $P(\text{P}) = \frac{2}{11}$

c) $P(\text{not S}) = \frac{7}{11}$

d) $P(\text{vowel}) = \frac{4}{11}$

e) $P(\text{P or I}) = \frac{6}{11}$

1) Mary chooses an integer at random from 1 to 6. What is the probability that the integer she chooses is a prime number?

2) A box contains six black balls and four white balls. What is the probability of selecting a black ball at random from the box?

3) A six-sided number cube has faces with the numbers 1 through 6 marked on it. What is the probability that a number less than 3 will occur on one toss of the number cube?

4) A fair coin is thrown in the air four times. If the coin lands with the head up on the first three tosses, what is the probability that the coin will land with the head up on the fourth toss?

5) What is the probability of choosing a king from a standard deck of playing cards?

6) What is the probability of choosing a green marble from a jar containing 5 red, 6 green, and 4 blue marbles? What is the probability of choosing a marble that is not blue?

7) What is the probability of getting an odd number when rolling a single 6-sided die?

8) What is the probability of choosing a jack or a queen from a standard deck of 52 playing cards?

SOLUTIONS

1) $\frac{3}{6}$ or $\frac{1}{2}$

2) $\frac{6}{10}$ or $\frac{3}{5}$

3) $\frac{2}{6}$ or $\frac{1}{3}$

4) $\frac{1}{2}$

5) $\frac{4}{52}$ or $\frac{1}{13}$

6) $\frac{6}{15}$ or $\frac{2}{5}$

$\frac{11}{15}$

7) $\frac{3}{6}$ or $\frac{1}{2}$

8) $\frac{8}{52}$ or $\frac{2}{13}$

16.4 How Do We Find the Probability of Compound Events?

When figuring out the probability of more than one event, you first need to determine if the events are dependent or independent.

Independent events   Two or more events in which the outcome of one event has no effect on the outcome of the other event or events.
If the events are independent, then you only need to multiply the probability of each event together.

Example:

A fair coin and a six-sided die are tossed simultaneously. What is the probability of obtaining a head on the coin and a 4 on the die jointly?

First, find the probability of a head on the coin

$$\frac{1}{2}$$

Then, find the probability of a 4 on the die

$$\frac{1}{6}$$

Multiply the two probabilities together to find the probability of both a head on the coin and a 4 on the die jointly.

$$\tfrac{1}{2} \times \tfrac{1}{6} \times \tfrac{1}{12}$$

Example:

A fair coin and a six-sided die are tossed simultaneously.

What is the probability of obtaining the following:

a head and a 3
$$\tfrac{1}{2} \times \tfrac{1}{6} = \tfrac{1}{12}$$

a head and an even number
$$\tfrac{1}{2} \times \tfrac{3}{6} = \tfrac{3}{12} \text{ or } \tfrac{1}{4}$$

a tail and a number less than 5
$$\tfrac{1}{2} \times \tfrac{4}{6} = \tfrac{4}{12} \text{ or } \tfrac{1}{3}$$

a tail and a number greater than 4
$$\tfrac{1}{2} \times \tfrac{2}{6} = \tfrac{2}{12} \text{ or } \tfrac{1}{6}$$

Dependent events Two events in which the outcome of the first event affects the outcome of the second event.

You need to find the probability of the first event. Then you assume the first event happened and find the probability of the second event based upon the first event being true.

Example:

A jar contains two red and five yellow marbles. A marble is drawn at random and not replaced. A second marble is drawn at random.

Find the probability that:

both marbles are red
$$\tfrac{2}{7} \times \tfrac{1}{6} = \tfrac{2}{42} \text{ or } \tfrac{1}{21}$$

Assume the one chosen was red, so now there is only 1 red out of the six remaining marbles.

both marbles are yellow
$$\tfrac{5}{7} \times \tfrac{4}{6} = \tfrac{20}{42} \text{ or } \tfrac{10}{21}$$

Assume the marble chosen was yellow. Now there are only 4 yellow marbles out of the six remaining.

Example:

Two cards are drawn from a deck of cards. The first card is drawn from the 52, not replaced, and then a second card is drawn.

$$P(\text{king, king})$$
$$\frac{4}{52} \times \frac{3}{51} = \frac{12}{2652} \text{ or } \frac{1}{221}$$

Assume the king was chosen, so now there are only 3 kings in the deck with 51 remaining cards.

$$P(\text{red, black})$$
$$\frac{26}{52} \times \frac{26}{51} = \frac{676}{2652} \text{ or } \frac{13}{51}$$

Assume a red card was chosen, so there are still 26 black cards; however, there are 51 cards left in the deck.

$$P(\text{heart, heart})$$
$$\frac{13}{52} \times \frac{12}{51} = \frac{156}{2652} \text{ or } \frac{1}{17}$$

Assume the heart was chosen, so now there are only 12 hearts in the deck with 51 remaining cards.

$$P(\text{heart, club})$$
$$\frac{13}{52} \times \frac{13}{51} = \frac{169}{2652} \text{ or } \frac{13}{204}$$

Assume a heart was chosen, so there are still 13 clubs; however; there are 51 cards left in the deck.

1) Two fair coins are tossed. What is the probability that both land heads up?

2) A jar contains four red and five yellow marbles. If one marble is drawn at random, the marble is replaced and a new marble is drawn. What is the probability of:

   a) (red, red)
   b) (yellow, red)
   c) (orange, yellow)

3) A jar contains four red and five yellow marbles. If one marble is drawn at random, the marble is not replaced and a new marble is drawn. What is the probability of:

   a) (red, red)
   b) (yellow, red)
   c) (orange, yellow)

4) Two cards are drawn from a standard deck. The first card is not put back before the second card is drawn. What is the probability of:

   a) $P(\text{ace, king})$
   b) $P(\text{red 5, black 5})$
   c) $P(\text{jack, jack})$

SOLUTIONS

1) $\frac{1}{2} \times \frac{1}{2} = \frac{1}{4}$

2) a) $\frac{4}{9} \times \frac{4}{9} = \frac{16}{81}$

   b) $\frac{5}{9} \times \frac{4}{9} = \frac{20}{81}$

   c) $\frac{0}{9} \times \frac{5}{9} = 0$

3) a) $\frac{4}{9} \times \frac{3}{8} = \frac{12}{72}$ or $\frac{1}{6}$

   b) $\frac{5}{9} \times \frac{4}{8} = \frac{20}{72}$ or $\frac{5}{18}$

   c) $\frac{0}{9} \times \frac{5}{8} = 0$

4) a) $\frac{4}{52} \times \frac{4}{51} = \frac{16}{2652}$ or $\frac{4}{663}$

   b) $\frac{2}{52} \times \frac{2}{51} = \frac{4}{2652}$ or $\frac{1}{663}$

   c) $\frac{4}{52} \times \frac{3}{51} = \frac{12}{2652}$ or $\frac{1}{221}$

16.5 What Is a Permutation?

An arrangement or listing in which order is important is called a **permutation.**

The symbol $P(6, 3)$ represents the number of permutations of 6 things taken 3 at a time. This can also be written as $_6P_3$.

Examples:

Maggie has 6 favorite ice cream flavors: How many three-scoop arrangements can she make on her cone?

$$6 \times 5 \times 4 = 120 \qquad 120 \text{ different combinations}$$

It is similar to the counting principle, where you multiply the choices together,

$_6P_3$ or $P(6, 3)$ means start with 6 and use 3 factors.

$$6 \times 5 \times 4 \qquad \text{Six things taken three at a time}$$

Beck has a CD player that allows the songs to play in a random order. He puts in a CD that has 12 songs. How many arrangements of 4 songs could randomly play?

12 songs to choose from, choosing 4 songs in order

$$_{12}P_4 = 12 \times 11 \times 10 \times 9 = 11{,}880$$

There are 11,880 arrangements.

Coach Eric is picking the first three batters out of nine players on the softball team. How many ways can he arrange the first three batters?

9 players to chose from, choosing 3 batters

$$9 \times 8 \times 7 = 504$$

There are 504 batting arrangements for the first three batters.

 EXAMPLE 16.5

1) A contractor can build 8 different model homes. He has 4 lots to build on. In how many ways can he put a different house on each lot?
2) There are seven teams in the MLS Eastern Conference. If there are no ties for placement in the conference, in how many ways can the teams place?
3) A security keypad system offers 10 digits on the keypad. If no digit is repeated, how many 4-digit codes can be made?
4) The school talent show has 15 acts. In how many ways can they award first, second, and third prize?

SOLUTIONS

1) $8 \times 7 \times 6 \times 5 = 1{,}680$
2) $7 \times 6 \times 5 \times 4 \times 3 \times 2 \times 1 = 5{,}040$
3) $10 \times 9 \times 8 \times 7 = 5{,}040$
4) $15 \times 14 \times 13 = 2{,}730$

16.6 What Is a Combination?

An arrangement or listing in which order is not important is called a **combination.**

The symbol $C(10,4)$ represents the number of combinations of 10 things taken 4 at a time. This can be written as $_{10}C_4$.

For combinations, the order is not going to be important. If there were 10 teams in the competition and the top 4 teams go on to the final round, how many different groups of four teams could be chosen to play in the final round?

You need to exclude the repetition of ways things can be arranged.

First, 10 things taken 4 at a time.

$$10 \times 9 \times 8 \times 7 = 5{,}040$$

Second, the different ways 4 teams can be arranged.

$$4 \times 3 \times 2 \times 1 = 24$$

Third, divide the 5,040 by the 24 to exclude the repetition.

$$\frac{5040}{24} = 210$$

There are 210 ways the top 4 teams can be chosen.

Think about it: If there are 4 people in a room, how many handshakes will occur if each person shakes hands with every other person?

A shakes *B*

A shakes *C*

A shakes *D*

B doesn't need to shake with *A*; they already shook hands

B shakes *C*

B shakes *D*

C doesn't need to shake with *A* or *B*; they already shook hands

C shakes *D*

D has already shaken everybody's hand

Therefore, there would be 6 arrangements of handshakes.

$$C(4,2) = \frac{4 \times 3}{2 \times 1} = 6$$

1) How many three-topping pizzas can be ordered from a selection of 12 pizza toppings?
2) How many committee members can be chosen for a five-person committee out of 30 candidates?
3) How many ways can you choose two co-captains for a soccer team from 18 players?
4) How many ways can you choose 3 shirts from the 7 in your closet?

SOLUTIONS

1) $C(12,3) = \dfrac{12 \times 11 \times 10}{3 \times 2 \times 1} = \dfrac{1320}{6} = 7,920$

2) $C(30,5) = \dfrac{30 \times 29 \times 28 \times 27 \times 26}{5 \times 4 \times 3 \times 2 \times 1} = \dfrac{17,100,720}{120} = 142,506$

3) $C(18,2) = \dfrac{18 \times 17}{2 \times 1} = \dfrac{306}{2} = 153$

4) $C(7,3) = \dfrac{7 \times 6 \times 5}{3 \times 2 \times 1} = 35$

Chapter Review

1) Determine if the situation is a permutation or a combination. Then find the number of possible outcomes.
   a) choosing a president and vice-president from a group of 20 members of a club
   b) choosing a committee of 4 from a group of 36
   c) choosing the starting 5 of a basketball team of 12 players
   d) choosing a three-digit code for a lock from 10 possible digits

2) You are at a carnival. One of the carnival games asks you to pick a door and then to pick a curtain behind the door. There are 3 doors with 4 curtains behind each door. How many choices are possible?

3) There are 3 trails leading to Camp A from your starting position. There are 3 trails from Camp A to Camp B. How many different routes are there from the starting position to Camp B? Draw a tree diagram to illustrate your answer.

4) A spinner has 4 equally likely regions numbered 1 to 4. Find the probability of each of the following:
   a) $P(1)$
   b) $P(2 \text{ or } 4)$
   c) $P(\text{odd})$
   d) $P(\text{whole number})$
   e) $P(\text{number greater than } 4)$

5) Using the same spinner, the arrow is spun twice. What is the probability that the spinner will land on a 1 on the first spin and a 2 on the second spin? Create a sample space to list the possibilities.

6) There are 2 red, 5 green, and 8 yellow marbles in a jar. Amanda randomly selects two marbles without replacing the first marble. What is the probability that she selects:
   a) two green marbles
   b) two red marbles
   c) a yellow and then a red

7) There are 2 red, 5 green, and 8 yellow marbles in a jar. Amanda randomly selects a marble, records the color, and then replaces the first marble before selecting another marble. What is the probability that she selects:
   a) two green marbles
   b) two red marbles
   c) a yellow and then a red

Statistics

WHAT YOU WILL LEARN

- To find mean, median, mode, and range.
- To understand center, spread, and overall shape of data.
- To interpret data in a bar graph, line graph, and circle graph.
- To display and interpret information in a dot plot.
- To display and interpret information in a stem-leaf plot.
- To display and interpret information in a box-whisker.
- To display information in a histogram.
- To display information in a scatter plot and investigate patterns of association in bivariate data.

<table>
<tr><td>

SECTIONS IN THIS CHAPTER

- What Are Statistics?
- What Are Measures of Central Tendency?
- How Do We Interpret Data?
- How Do We Create Dot Plots?
- How Do We Create a Stem-Leaf Plot?
- How Do We Create a Box-Whisker?
- How Do We Create a Histogram?
- How Do We Create a Scatter Plot?

</td></tr>
</table>

17.1 What Are Statistics?

Statistics is the collection, organization, presentation, and analysis of data. Have you ever been home when the phone rang and someone just wanted to ask you a few questions? This is a person conducting a survey. A survey is when you are asked either written or verbal questions for the purpose of acquiring information/data. This is often done for elections. The candidates use these polls to help plan their strategy. Statistics are used quite often in decision making in politics, education, and business.

First, you gather the information. This can be done by survey, poll, sampling, or study. Sometimes data can be obtained from every member of a group—for example, to find out what everyone in the class wants as a class pet. However, it is often impossible to ask everyone; that is when a sampling would be done.

Population   A group of people, objects, or events that fit a particular description; in statistics, the set from which a sample of data is selected.

Random sample   A sample obtained by a selection from a population, in which every element of the population has an equal chance of being selected.

Sample   A representative part or a single item from a larger whole or group; a finite part of a statistical population whose properties are studied to gain information about the whole.

Sampling   Selecting a small group which is representative of the entire population; used in taking a survey.

Second, you organize the data. The organization generally starts with a chart and is then based on the need and type of data a graph would be chosen to represent.

Third, you share the data with others. This is when your graphs are very important. You always want to choose an appropriate graph.

These are some graphs you should already be familiar with:

Bar graph   A graph that uses horizontal or vertical bars to display data.

Example:

Circle graph or pie graph   A graph in which the data is represented by sectors of a circle; the total of all the sectors should be 100% of the data.

Example:

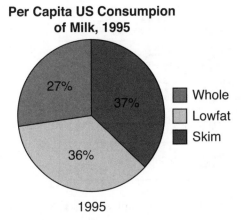

Per Capita US Consumpion of Milk, 1995

27%
37%
36%

■ Whole
□ Lowfat
■ Skim

1995

Double-bar graph   A graph that uses pairs of bars to compare and show the relationship between data.

Double-line graph   A graph that uses pairs of lines that show change over time to represent and compare data.

Line graph   A graph that uses line segments to show changes in data; the data usually represents a quantity changing over time.

Example:

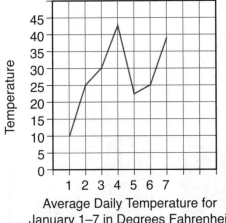

Average Daily Temperature for January 1–7 in Degrees Fahrenheit

Pictograph   A graph that uses pictures or symbols to represent data; an accompanying key indicates the value associated with each picture or symbol.

Example:

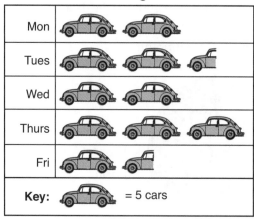

Cars Sold During One Week

Mon	(2 cars)
Tues	(2½ cars)
Wed	(2 cars)
Thurs	(3 cars)
Fri	(1½ cars)
Key:	= 5 cars

Fourth, analyze the data. Often, misleading graphs are used to persuade people. Skill in understanding and interpreting data is vital in evaluating news broadcasts, advertisements, and political campaigns. Two different candidates can display the same data and mislead the consumer or viewer toward their point of view.

Misleading graph   A graph that leads the reader to make an incorrect conclusion or to form a false impression.

Let's look at some graphs.

Bar Graphs
Single Bar Graph

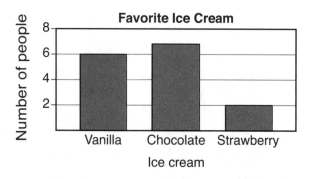

Favorite Ice Cream

There is a lot of information you can gather from the graph.

Which is the favorite ice cream? chocolate

Which is the least favorite ice cream? strawberry

How many people like each flavor the best? vanilla: 6, chocolate: 7, strawberry: 2

How many more people like chocolate than strawberry? $7 - 2 = 5$; 5 more people like chocolate.

Double Bar Graph

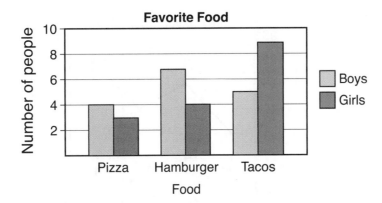

The graph shows the same information for two groups: boys and girls.

What is the favorite boy's food? hamburgers

What is the favorite girl's food? tacos

What is the least favorite boy's food? pizza

What is the least favorite girl's food? pizza

How many like each food? pizza 4 + 3 = 7, hamburgers 7 + 4 = 11, tacos 5 + 9 = 14

How many more boys like pizza than girls? 4 − 3 = 1, 1 more boy likes pizza

Circle Graphs

The graph is used to show parts of a whole. If you do not know the number of people surveyed, you cannot answer number-specific questions like "how many"; however, you can answer questions with percentages.

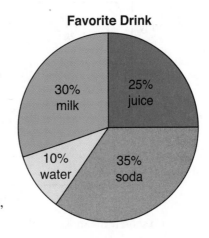

Favorite Drink

Which is the favorite drink? soda

Which is the least favorite drink? water

What percent like each drink the best? soda 35%, juice 25%, milk 30%, water 10%.

Line Graphs

Which month did they save the most? April

Which month did they save the least? January

How much did they save each month? January – 5, February – 15, March – 10, April – 20.

Picture Graphs

Picture Graphs show the same information as a bar graph with pictures instead of bars.

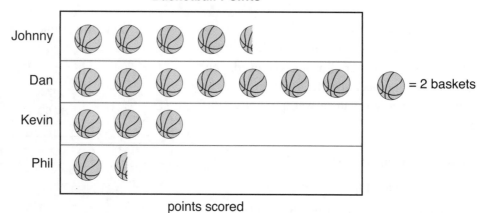

Who scored the most points? Dan

Who scored the least points? Phil

How many more points did Dan score than Johnny? 5 points

EXAMPLE
17.1

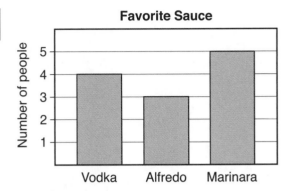

Favorite Sauce

1) What is the favorite sauce?
2) What is the least favorite sauce?
3) How many people were surveyed?

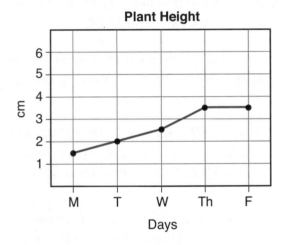

Plant Height

4) At what height did the plant start?
5) How tall was the plant on Wednesday?
6) Which day did the plant grow the most?

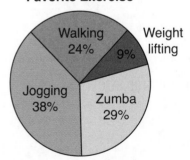

Favorite Exercise

7) What is the favorite exercise?
8) What is the least favorite exercise?
9) What percentage like zumba?

SOLUTIONS

1) Marinara
2) Alfredo
3) 12
4) 1.5 cm
5) 2.5 cm
6) Thursday
7) jogging
8) weightlifting
9) 29%

17.2 What Are Measures of Central Tendency?

A single number that represents a typical value for a set of numbers; the three most common measures of central tendency are the mean, median, and mode.

Mean A measure of central tendency; the quotient obtained when the sum of the numbers in a set is divided by the number of addends.

Example:

Karli took four exams in chemistry. Her grades were 86, 92, 84, and 78. To calculate her average, you first add all her grades together: $86 + 92 + 84 + 78 = 340$. Next, you divide by the number of tests: $\frac{340}{4} = 85$. Her mean is 85.

Median The middle number of a set of numbers arranged in increasing or decreasing order; if there is no middle number, the median is the average of the two middle numbers.

Example:

All the students were jumping to see how high they could reach. They reached the following heights in cm:

120, 135, 147, 154, 187, 191, 198, 199, 210

187 is the middle number; it is the median

Another student came in late and wanted to try. He jumped 165 cm.

120, 135, 147, 154, 165, 187, 191, 198, 199, 210

Add $165 + 187$

Divide $\frac{352}{2} = 176$

176 is the new median

Mode   The number or members of a data set that occurs most frequently in the set. There can be no mode, one mode, or even two modes. If there are more than two numbers that appear the same amount of times, we generally say that there is no mode.

Example:

Chris recorded the age of customers at the video game store for two hours. The ages were:

12, 12, 13, 13, 14, 14, 14, 14, 16, 16, 18, 19, 21, 22, 22, 23, 42, 44, 45, 48

The mode is 14 as there were four customers at that age.

Range of a data set   The difference between the greatest and the least values in a set of numbers.

Example:

Tom recorded the tips he received during his breakfast deliveries. He received:

$1.25, $2.35, $3.10, $3.18, $3.75, $4.15, $5.00

The lowest amount was $1.25. The highest amount was $5.00. The difference between the two is $5.00 − $1.25 = $3.75. The range of tips is $3.75.

1) Find the mean, median, mode, and range of each set of data.
   a) 20, 21, 15, 19, 17, 15, 14, 12, 16, 13, 18, 15
   b) 132, 140, 140, 149, 150, 151, 153, 154, 155, 160, 162, 172
   c) 601, 461, 431, 400, 357, 330, 313, 309, 306, 290

SOLUTIONS

1) a) Mean = 16.25
      Median = 15.5
      Mode = 15
      Range = 9
   b) Mean = 151.5
      Median = 152
      Mode = 140
      Range = 40
   c) Mean = 379.8
      Median = 343.5
      Mode = none
      Range = 311

17.3 How Do We Interpret Data?

Data is often organized using tables and represented visually with different graphs. The graph is generally matched up with the purpose or with the intended audience. Sometimes, bar graphs will sometimes have background images to make them more interesting in a presentation.

Sometimes part of the data is not represented. For example, an outlier may be removed from the data set if it is irrelevant or seen as an error. If we collected data on the height of second graders and one of the students was recorded as being 62 inches tall, the likelihood of that being a correct measurement is small, so we could exclude the data point.

Let's look at a sample of graphs and answer some questions.

Bar Graphs

A bar graph is used for comparisons.

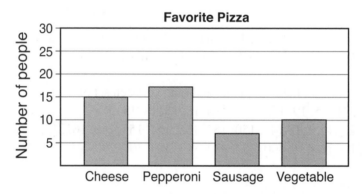

When analyzing a graph, you are making predictions based on the gathered information. For example, if you owned a pizza shop, you would be able to predict sales. You would estimate how many of each type of pizza to make. They should make more pepperoni than any other pizza. If they were having a party, they could predict to make a ratio of 4 pepperoni, 3 cheese, 1 sausage, and 2 vegetable pizzas by using estimation.

Line Graphs

A line graph is used to show trends or change over time.

A line graph is read left to right.

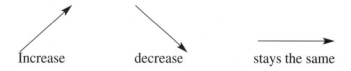

A person was walking to school; the graph below shows their journey.

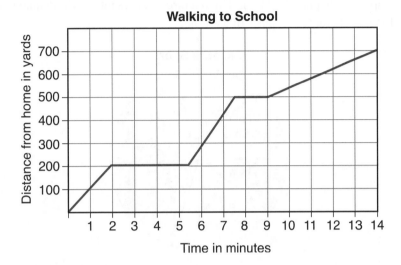

Walking to School

The steeper lines show a faster rate. The horizontal line shows that time moved, but the distance remained the same, so the person is standing still. We can infer reasons as to why someone would stand still. Perhaps they were waiting for a car to pass or waiting to meet up with a friend.

Picture Graphs

In picture graphs, symbols are used. They show the same information as a bar graph, but in a more interesting way. Magazines usually prefer a picture graph.

Super Bowl Wins

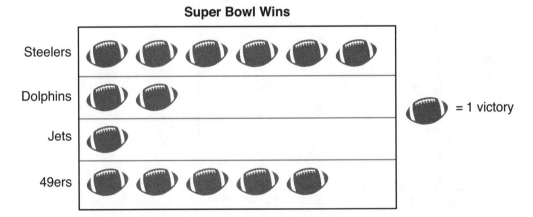

Picture graphs can be used in the same way as a bar graph to make predictions. If it was known that Super Bowl wins are proportionate to the number of fans, we would predict that the Steelers have the most fans. If there are 400 football fans of the four teams, the Steelers have 6 wins out of a total of 14. We would expect $\frac{6}{14}$ to be the same as around 171 out of 400 by using a proportion.

Circle Graphs

A circle graph is a display of data in which a circle represents a whole quantity, and sectors of the circle represent parts of a whole.

Favorite Sport

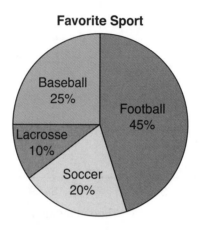

We can use the data to predict. If a restaurant was running specials on game day, they would expect more people attending the football special. You can also use the percentage to do calculations. If there are 250 people, how many would be soccer fans?

$$250 \times .2 = 50 \qquad 50 \text{ people}$$

EXAMPLE 17.3

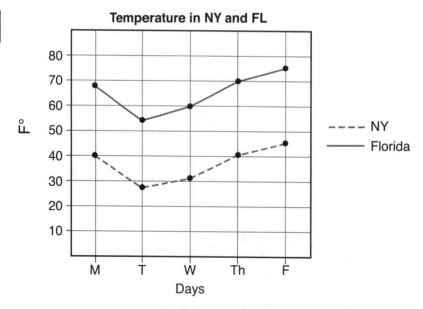

1) How much warmer is Florida than New York on Monday?
2) What is a trend both cities are having from Wednesday to Friday?
3) In which state on which day might it snow?

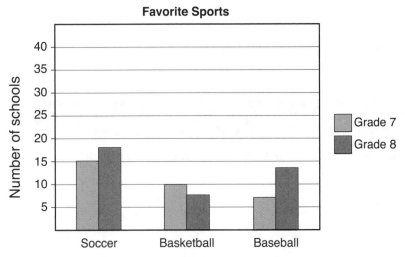

Favorite Sports

4) What is the most popular sport?

5) If you were making favors for a party based on sports for 200 kids, how many of baseball would you make?

6) Which sport do more 7th graders prefer compared to 8th graders?

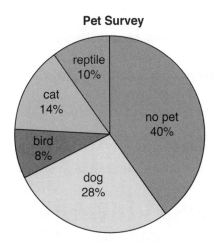

Pet Survey

7) Which pet do more students have?

8) Would asking students to draw their pet be a good class activity?

9) If there were 500 students in the school, about how many students would you predict have birds as pets?

SOLUTIONS

1) $68 - 40 = 28$; 28 degrees warmer in Florida

2) In both cities, it is getting warmer each day.

3) It would more likely snow in New York on Tuesday because it is below freezing.

4) Soccer

5) Soccer ($15 + 18 = 33$); basketball ($10 + 8 = 18$); baseball ($7 + 14 = 21$)
Total is 72. Baseball is $\frac{21}{72}$. Set up a proportion: $\frac{21}{72} = \frac{x}{200}$ around 58.

6) More 7th graders prefer basketball than 8th graders.
7) Dogs
8) No, it would not be a good activity, since 40% of the students do not have a pet.
9) 8% of 500 is 40; we would predict 40 students.

17.4 How Do We Create Dot Plots?

A dot plot is a line plot where you show data on a number line with dots to show frequency.

Below is an example of a dot plot.

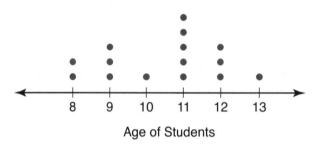

Age of Students

The count of each dot above the score represents the number of students who received the score.

To create a dot plot, set up a number line and place a dot for each value to show the frequency.

For example, the price of candles from 15 different stores is listed below.

$10, $12, $8, $7, $10, $12, $13, $15, $12, $8, $9, $12, $8, $6, $15

To create a dot plot, it is helpful to put your data in order from least to greatest so you can set up an appropriate scale on the number line.

$6, $7, $8, $8, $8, $9, $10, $10, $12, $12, $12, $12, $13, $15, $15

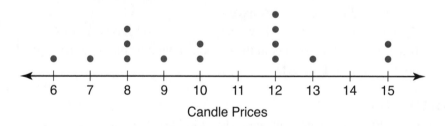

Candle Prices

The data points that are the same will be grouped on top of one another like a stack.

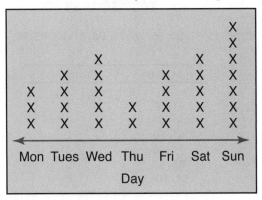

Another example, we will create a line plot to show the heights in inches of 20 girls in a karate club.

46, 46, 47, 48, 48, 48, 50, 51, 51, 54, 54, 55, 56, 57, 57, 57, 57, 57, 58, 60

Set up the number line.
Place a dot for each data point.

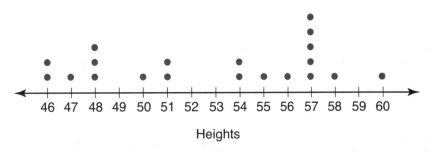

Heights

The number line does not have to have only numbers as labels; you can use almost any data you are given. For example, the dot plot below shows the days of the week on the number line.

Number of Hours Spent Watching TV

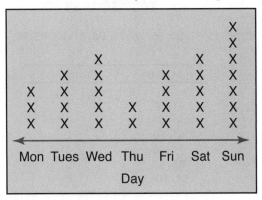

EXAMPLE
17.4

Create a dot plot for each situation.
1) Daniel has a collection of football jerseys containing the following numbers.
12, 10, 18, 22, 10, 26, 18, 10, 34, 10, 42, 10, 44, 44
2) Elisa recorded all her test scores for all her classes during the third grading period.
84, 87, 92, 93, 94, 87, 94, 95, 96, 94, 88, 90, 92, 95, 94, 97, 98, 95, 93

SOLUTIONS

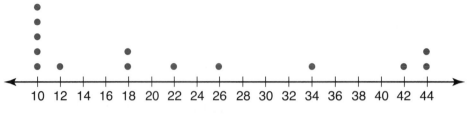

Daniel's Jerseys

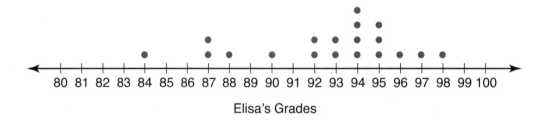

Elisa's Grades

17.5 How Do We Create a Stem-Leaf Plot?

There are a variety of ways to organize data. You can group data using intervals in a table. This works well; however, you lose the individual data points.

TEST SCORES

84, 87, 92, 67, 45, 87, 90, 95, 76, 78, 68, 81, 83, 95, 99, 76, 84, 97, 71

Intervals	Frequency
41–50	1
51–60	0
61–70	2
71–80	4
81–90	6
91–100	6

You can create a stem-leaf plot. This allows you to group data without losing individual data points. You start by deciding what place value will be the stem, and then arrange your leaves in order.

TEST SCORES

```
4 | 5
5 |
6 | 7  8
7 | 1  6  6  8
8 | 1  3  4  4  7  7
9 | 0  2  5  5  7  9
```

Key 4|5 = 45

A stem-leaf plot is used to organize large data sets in a small amount of space. In stem-leaf plots, the numerical data are listed in ascending or descending order. The digits with the greatest common place value are used for the stems. The digits in the next greatest place value are used for the leaves.

At the family reunion all the relatives recorded their age on a poster.

The ages were 2, 3, 5, 5, 11, 12, 12, 13, 15, 16, 21, 24, 26, 39, 44, 45, 48, 49, 65, 67, and 70.

First, identify the smallest and the largest numbers.

2 is the smallest.
70 is the largest.
The greatest place value is the tens place.

Second, draw a vertical line and write the stems to the left of the line in order.

```
0 |
1 |
2 |
3 |
4 |
5 |
6 |
7 |
```

Third, write the leaves to the right of the corresponding stems on the other side of the vertical line. For example, 70 would be written by adding a 0 to the right of the 7 in the tens place.

FAMILY REUNION

```
0 | 2  3  5  5
1 | 1  2  2  3  5  6
2 | 1  4  6
3 | 9
4 | 4  5  8  9
5 |
6 | 5  7
7 | 0
```

Lastly, make sure to include a key.   1|1 = 11

1) Create a stem-leaf plot using the given information.
   The age of U.S. Presidents at time of inaugurations.
   42, 43, 46, 46, 47, 47, 48, 49, 49, 50, 51, 51, 51, 51, 51,
   52, 52, 54, 54, 54, 54, 54, 55, 55, 55, 55, 56, 56, 56, 57
   57, 57, 57, 58, 60, 61, 61, 61, 62, 64, 64, 65, 68, 69

2) Create a stem-leaf plot using the given information.
   The weight of students in Period 9
   95, 103, 106, 110, 112, 112, 115, 115, 118, 119, 120, 124, 127, 132, 135, 160, 165

SOLUTIONS

1) 4 | 2 3 6 6 7 7 8 9 9
    5 | 0 1 1 1 1 1 2 2 4 4 4 4 4 5 5 5 5 6 6 6 7 7 7 7 8
    6 | 0 1 1 1 2 4 4 5 8 9

   Key 4|2 = 42

2) 9 | 5
    10 | 3   6
    11 | 0   2   2   5   5   8   9
    12 | 0   4   7
    13 | 2   5
    14 |
    15 |
    16 | 0   5

   Key 11|2 = 112

17.6 How Do We Create a Box-Whisker?

A box-whisker uses a number line to show the distribution of a set of data by using five values. The box is drawn around the middle half of the data, and the whiskers extend from each quartile to the extreme data points that are not outliers. A vertical line is drawn through the box at the median.

The box-whisker is constructed by finding the median, then finding the medians of each half of data. This divides the data into quartiles.

Let's look at an example:

Elevation of Selected U.S. cities

92, 114, 130, 158, 162, 164, 193, 209, 330

Median

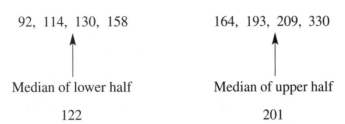

Median of lower half   Median of upper half

122   201

Now, we have our five data points:

   92 – Minimum
  122 – First quartile
  162 – Median or second quartile
  201 – Third quartile
  330 – Maximum

Set up a number line.

80   100   120   140   160   180   200   220   240   260   280   300   320   340

Draw in the box-whisker.
Set up the box using the quartiles.
Add the whiskers using the extremes.

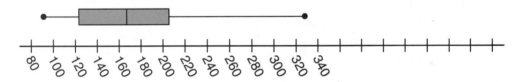

The box-whisker plots separate data into four parts. Although the parts usually differ in length, each part contains one-fourth of the data.

The size of the whiskers shows the size of the range. Longer whiskers indicate a greater range. Shorter whiskers indicate a lesser range.

Outliers can be shown by adding a separate data point with a star or a point.

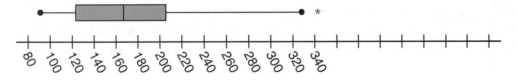

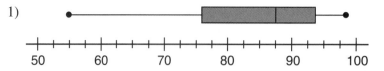

EXAMPLE 17.6

1)

Find the minimum, first quartile, median, third quartile, maximum, and range from the box-whisker above.

2) Construct a box-whisker from the given data.
   Height in inches of 8th graders
   54, 56, 59, 60, 61, 61, 62, 62, 62, 63, 65, 66,
   66, 67, 67, 68, 72

SOLUTIONS

1) 55 – Minimum
    76 – First quartile
    87 – Median or second quartile
    93 – Third quartile
    98 – Maximum

2)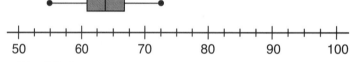

    54 – Minimum
    60.5 – First quartile
    62 – Median or second quartile
    66.5 – Third quartile
    72 – Maximum

17.7 How Do We Create a Histogram?

Frequency table   A table that shows how often each item, number, or range of numbers occurs in a set of data.

Histogram   A special kind of bar graph that displays the frequency of data that has been organized into equal intervals; the intervals cover all possible values of data, therefore there are no spaces between the bars of the graph; the horizontal axis is divided into continuous equal intervals.

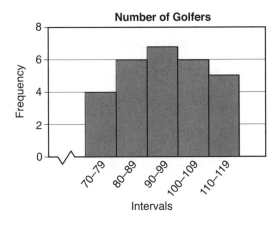

TEST SCORES

51, 56, 65, 67, 68, 70, 73, 81, 82, 83, 85, 85, 90, 91, 94, 97, 99

First, organize your data by deciding on the size of the intervals. All the intervals must be of equal size. You want to have around 5–10 intervals, so deciding on the interval range is important. Since our lowest number is 51 and the highest is 99, if we use an interval size of 10 that would work well for this data.

Next, complete the chart.

Intervals	Frequency
51–60	2
61–70	3
71–80	2
81–90	5
91–100	5

Next, construct the histogram. The vertical axis is the frequency. The frequency scale always begins with zero. The horizontal axis would be the intervals. Arrange these equal intervals in order, from least to greatest. You would want to leave a gap from the origin since your starting the lowest interval at 51.

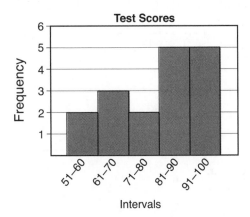

Try another:

Record of temperatures in degrees Fahrenheit for November

31, 32, 33, 38, 42, 42, 44, 44, 45, 45, 46, 47, 52, 53, 53, 54, 54, 54, 55, 56, 57, 59

Decide on the interval. An interval of 5 would work well with this data.

Next, complete the chart.

Intervals	Frequency
31–35	3
36–40	1
41–45	6
46–50	2
51–55	7
56–60	3

Construct the histogram.

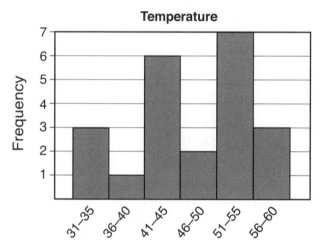

 1) Test scores
   51, 51, 54, 62, 64, 72, 72, 75, 77, 78, 81, 81, 82, 85, 86, 88, 91, 92, 94, 95, 98
   Complete the chart.

Intervals	Frequency
51–60	
61–70	
71–80	
81–90	
91–100	

Construct the histogram.

SOLUTIONS

1)

Intervals	Frequency
51–60	3
61–70	2
71–80	5
81–90	6
91–100	5

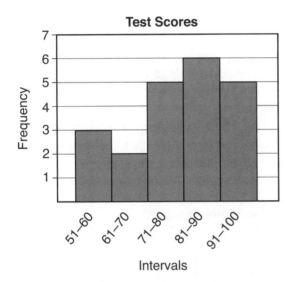

17.8 How Do We Create a Scatter Plot?

Scatter plots are similar to line graphs in that they use horizontal and vertical axes to plot data points. However, they have a very specific purpose. Scatter plots show how much one variable is affected by another. If there are two variables, the data is considered bivariate. The relationship between two variables is called their **correlation**.

Scatter plots usually consist of a large body of data. The closer the data points come when plotted to making a straight line, the higher the correlation between the two variables, or the stronger the relationship. The further apart the data points, the weaker the correlation. If there is a weak correlation, the variables are not affected by one another. There can be a weak correlation, or none at all.

If the data points make a straight line going from the origin out to high x- and y-values, then the variables are said to have a **positive correlation**. If the line goes from a high value on the y-axis down to a high value on the x-axis, the variables have a **negative correlation**. Also, sometimes there is a correlation but no causation, which means one did not cause the other to occur.

To make a scatter plot, you need to set up both x and y axes. Then plot the points as if you were plotting on the Cartesian plane.

Years of experience	Salary per hour
1	$10
2	$10
2	$11
2	$25
3	$12
3	$14
4	$14
5	$15
6	$18
6	$19
8	$20
8	$22
9	$23
10	$25

You need to decide on your *x* and *y* scale. The *x*-values range from 1 to 10, therefore a scale of 1 will work. The *y*-values go from 10 to 25, therefore a scale of 2 or 3 would work nicely. Once your axes are set, plot each point as if you had an (*x*, *y*). It is not necessary to label the points.

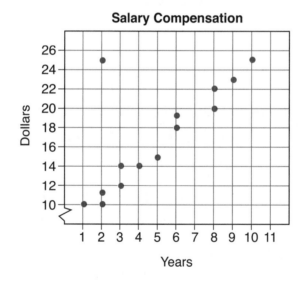

This graph shows a strong positive correlation except for one outlier. Therefore, we can conclude the hourly wage increases with the number of years experience.

Another example,

Saturday's temperature (in °F) and ice cream cone sales									
40	42	45	56	58	64	66	74	57	70
2	4	4	8	10	13	19	21	18	25

You need to decide on your *x* and *y* scale. The *x*-values range from 40 to 70, therefore a scale of 5 or 10 will work. The *y*-values go from 2 to 25, therefore a scale of 2 or 3 would work nicely. Once your axes are set, plot each point as if you had an (*x*, *y*). It is not necessary to label the points.

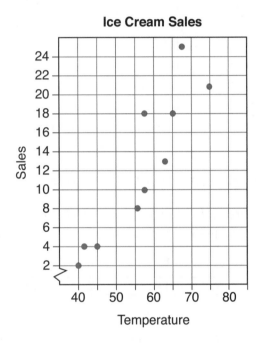

The scatter plot shows a correlation. It is not very strong—we would need more data—but it does show a positive correlation.

Another example:

Student height (in inches) and quiz scores												
60	61	62	63	63	65	66	68	68	70	71	71	72
92	87	76	89	90	97	98	90	87	82	85	90	79

You need to decide on your *x* and *y* scales. The *x*-values range from 60 to 72, therefore placing a break would be a good idea. A break allows the scale to start at a number other than 0. After the break, start with 60 and use a scale of 1. The *y*-values go from 76 to 98, therefore a break would be a good idea as well. After the break, starting at 76 and using a scale of 2 or 3 would work nicely. Once your axes are set, plot each point as if you had an (*x*, *y*). It is not necessary to label the points.

Heights Compared to Quiz Scores

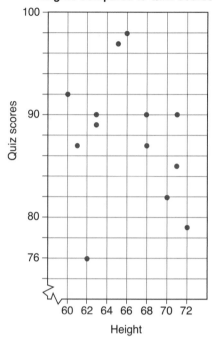

The points are all scattered, therefore there is no correlation.

EXAMPLE 17.8

Create a scatter plot and decide if there is a correlation.

1) Create a scatter plot to show the number of people at the party and the total cost in dollars.

Number of People	Cost (dollars)
12	200
17	288
24	400
30	550
36	612
42	700

SOLUTION

1)

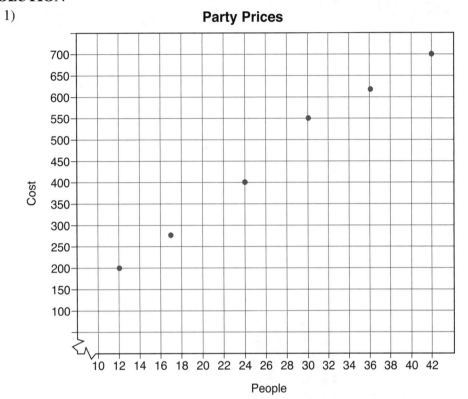

Chapter Review

1) On four Spanish tests Michelle received these grades:
   82   89   78   86
   What grade must Michelle get on her next test for her average to be an 85?

2) Find the mean, median, mode, and range for each set of data.
   a) 6   8   8   7   8   5
   b) 310   290   302   296   342   326   345   412   248

3) How many more students like chocolate than vanilla?

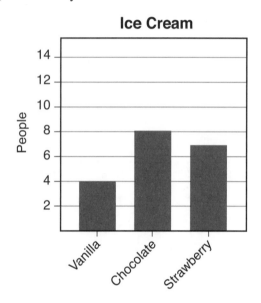

Ice Cream

4) How many centimeters did the plant grow from Week 3 to Week 7?

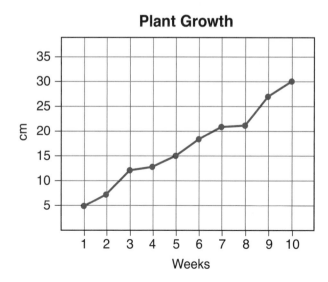

Plant Growth

5) Daily round-trip mileages for a group of commuters:

3	4	6	9	10	12	15	16	20	26
29	30	32	33	38	39	40	40	42	44

   Construct a stem-leaf plot and a box-whisker.

6) Points scored in a basketball game:

17	18	22	25	26	31	36	36	38	39
41	42	44	45	51	51	52	58	59	59

   Construct a histogram.

7) Maggie counted the number of students who had birthdays each month.

Jan	Feb	Mar	Apr	May	June	July	Aug	Sept	Oct	Nov	Dec
3	4	2	5	1	6	0	3	6	3	2	4

   Create a dot plot.

8) Create a scatter plot. Consider Monday as Day One.

   Test Scores and Day Exam Given

Monday	78, 92, 81
Tuesday	82, 95, 97
Wednesday	86, 92, 96
Thursday	81, 93, 95
Friday	86, 92, 98

9) The scatter plot below shows the appliances and the cost of the repairs. Is there a correlation? If so, which type of correlation?

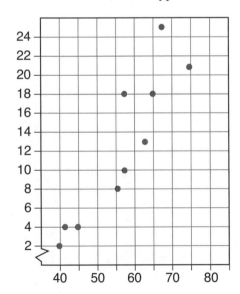

Sequences

WHAT YOU WILL LEARN

- How to find the terms of arithmetic sequences.
- How to find the terms of geometric sequences.

SECTIONS IN THIS CHAPTER
• What Are Patterns?
• What Are Arithmetic Sequences?
• What Are Geometric Sequences?

18.1 What Are Patterns?

Numeric pattern   An arrangement of numbers that repeat or follow a specified rule.

Pattern   A design (geometric) or sequence (numeric or algebraic) that is predictable because some aspect of it repeats.

It is only natural for us to look for patterns in things. Nature is full of patterns.

The triangle numbers are numbers generated from a pattern of dots which form a triangle.

The pattern looks like this:

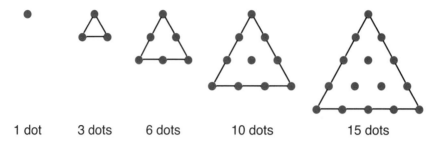

| 1 dot | 3 dots | 6 dots | 10 dots | 15 dots |

The numbers start 1, 3, 6, 10, 15, 21, 28, 36, ...

The square numbers are numbers generated from a pattern of dots which form squares. You can also find them by squaring.

$$1^2 = 1$$
$$2^2 = 4$$
$$3^2 = 9$$
$$4^2 = 16$$

The numbers start 1, 4, 9, 16, 25, 36, 49, ...

There are many other patterns.

Cube Numbers: 1, 8, 27, 64, 125, 216, 343, 512, ...

The next number is found by cubing.

Fibonacci numbers: 0, 1, 1, 2, 3, 5, 8, 13, 21, 34, ...

The next number is found by adding the two preceding numbers together.

Some patterns are sequences, a set of things in order. The things in the sequence are called members or elements. If the sequence goes on forever, it is called an infinite sequence; otherwise, it is called a finite sequence.

Sequences also use the same notation as sets use: list each element, separated by a comma, and then put brackets around the whole thing.

Example:
Even whole numbers

{ 0, 2, 4, 6, 8, 10, ...}   an infinite set

A sequence usually has a rule, which is a way of finding the value of each term. For ‸mple, we found the even whole numbers by starting with zero and adding two to ‸receding term.

Let's look at some sequences to try to find the pattern and the rule.

Examples:

$$\{1, 3, 7, 15, 31, ...\}$$

We know the numbers are increasing. The first increase is by 2, then by 4, then 8, and then 16, so we can generalize and say that the next increase will be by 32.

We can also look at any multiplication or addition that may have occurred.

$$2(1) + 1 = 3$$
$$2(3) + 1 = 7$$
$$2(7) + 1 = 15$$

It appears the pattern rule of $2n + 1$ would work as well.

People use patterns to predict. Patterns are analyzed to solve crimes as well. Patterns are used in designs for homes, clothes, and much more. Patterns can be very useful.

18.2 What Are Arithmetic Sequences?

A *sequence* is an ordered list of numbers. An ***arithmetic sequence*** is a sequence in which the difference between any two consecutive terms is the same. So, you can find the next term in the sequence by adding the same number to the previous term. Each number is called a term in the sequence.

$$5, 10, 15, 20, 25, 30, ...$$

The difference is called the ***common difference***. In this case the common difference is $+5$.

Sometimes you are asked to verify if a sequence is arithmetic. You would compare the terms and look for the common difference. If there is no common difference, then the sequence is not arithmetic.

Example:

$$8, 5, 2, -1, -4, ..$$

The common difference is -3, therefore it is an arithmetic sequence.

$$9, 4, 1, 0, ...$$

There is no common difference, therefore it is not an arithmetic sequence.

You can use the common difference to help you find terms in the sequence by continuing the pattern.

Using the sequence 8, 5, 2, −1, −4 we can find the next few terms by knowing the common difference is −3.

$$-4 - 3 = -7$$
$$-7 - 3 = -10$$
$$-10 - 3 = -13$$

EXAMPLE 18.2

1) State whether the sequence is arithmetic. If the sequence is arithmetic, name the common difference and the next three terms.
   a) 2, 5, 8, 11, ...
   b) −6, 5, 16, 27, ...
   c) $\frac{1}{2}$, 1, 2, 4, ...
   d) 2, 6, 18, 24,
   e) 18, 11, 4, −3,

SOLUTIONS

1) a) arithmetic
      Common difference: +3
      Next three terms: 14, 17, 20
   b) arithmetic
      Common difference: +11
      Next three terms: 38, 49, 60
   c) not arithmetic
   d) not arithmetic
   e) arithmetic
      Common difference: −7
      Next three terms: −10, −17, −24

18.3 What Are Geometric Sequences?

A *geometric sequence* is a sequence in which the quotient of any two terms is the same. So, you can find the next term by multiplying the previous term by the same number.

Each number is called a term in the sequence.

$$1, 5, 25, 125, 625, ...$$

The quotient is called the *common ratio*. In this case the common ratio is +5.

Sometimes you are asked to verify if a sequence is geometric. You would compare the terms and look for the common ratio. If there is no common ratio, then the sequence is not geometric.

Example:

$$4, -12, 36, -108, 324, ...$$

The common ratio is -3, therefore it is a geometric sequence.

$$1, 3, 6, 9, ...$$

There is no common ratio, therefore it is not a geometric sequence.

You can use the common ratio to help you find terms in the sequence by continuing the pattern.

Using the sequence 1, 2, 4, 8, 16, we can find the next few terms by knowing the common ratio is 2.

$$16 \times 2 = 32$$
$$32 \times 2 = 64$$
$$64 \times 2 = 128$$

 EXAMPLE 18.3

1) State whether the sequence is geometric. If the sequence is geometric, name the common ratio and the next three terms.
   a) 2, 6, 18, 54, ...
   b) 25, 22, 19, 16, ...
   c) 64, 32, 16, 8, ...
   d) 25, 5, 1, $\frac{1}{5}$, ...
   e) $-18, 6, -2, \frac{2}{3}$

SOLUTIONS

1) a) geometric
      Common ratio: 3
      Next three terms: 162, 486, 1458
   b) not geometric
   c) geometric
      Common ratio: $\frac{1}{2}$
      Next three terms: 4, 2, 1
   d) geometric
      Common ratio: $\frac{1}{5}$
      Next three terms: $\frac{1}{25}, \frac{1}{125}, \frac{1}{625}$
   e) geometric
      Common ratio: $-\frac{1}{3}$
      Next three terms: $-\frac{2}{9}, \frac{2}{27}, -\frac{2}{81}$

Chapter Review

1) Tell whether the sequence is arithmetic, geometric, or neither.

   a) 1,   3,   9,   27,   81

   b) 6,   8,   10,   12,   14

   c) 0,   5,   10,   15,   20

   d) 1,   1,   2,   3,   5

   e) 100,   10,   1,   .1,   .01

Cumulative Review

1) Find the percent increase of a stock that was originally $250 and is now $400.

2) Find the percent error (to the nearest percent) when the actual weight is 3.5 grams and you measured 3.65 grams.

3) You bought a shirt for $60 and paid 8% sales tax. What is the final cost?

4) Reflection over the y-axis
   A $(2, 2) \rightarrow A'$ (,)
   B $(4, 2) \rightarrow B'$ (,)
   C $(4, -4) \rightarrow C'$ (,)

5) Reflection over the x-axis
   A $(2, 2) \rightarrow A'$ (,)
   B $(4, 2) \rightarrow B'$ (,)
   C $(4, -4) \rightarrow C'$ (,)

6) Translate 7 units right and 4 units up
   A $(-4, -1) \rightarrow A'$ (,)
   B $(-4, -3) \rightarrow B'$ (,)
   C $(-1, -1) \rightarrow C'$ (,)

7) Reflection over the line $x = 1$
   A $(4, -1) \rightarrow A'$ (,)
   B $(4, -3) \rightarrow B'$ (,)
   C $(-1, -1) \rightarrow C'$ (,)

8) Make a box-and-whisker for the data below.

Temperature in Virginia over 9 months								
High	43	47	56	54	48	67	72	76	62
Low	32	34	43	51	53	47	42	31	24

9) Write an algebraic expression, equation, or inequality.
   a) A number is more than 2 and less than 8.
   b) The product of some number and 8, decreased by 2.
   c) A number decreased by the quotient of 8 and 2.
   d) Twice the sum of some number and 8.
   e) A number decreased by 8 is less than 2.
   f) The sum of a number and 2 is 8.
   g) 8 more than the sum of some number and 2.
   h) 8 less than some number is 2.

10) Solve each equation.
   a) $x + 8 = 12$
   b) $x - 4 = -8$
   c) $2x - 3 = 13$
   d) $3x = 12$
   e) $\frac{x}{3} - 6 = -1$
   f) $2(x - 14) = 18$
   g) $2x - 4x + 8 = -6$
   h) $2x - 8 = 4x + 12$
   i) $3x + x + 4 = 16$

11) Match each word phrase to a symbol.
   a) is equal to   1) $<$
   b) is not equal to   2) $=$
   c) is greater than   3) $>$
   d) is less than   4) $\leq$
   e) is greater than or equal to   5) $\neq$
   f) is less than or equal to   6) $\geq$

12) Represent the solution set on a number line.
   a) $x \leq -1$
   b) $3x - 8 > 13$
   c) $4x - 7x + 5 > 17$
   d) $12 > 2x + 6$

13) Evaluate:

    a) $216^{\frac{1}{3}}$

    b) 5^{-3}

    c) $16^{\frac{3}{2}}$

14) Find the missing side of the right triangle.

    a) $a = 5$ $b = 12$ $c = ?$

    b) $a = 14$ $b = ?$ $c = 50$

15)

    a) What is the favorite ice cream?

    b) What is the least favorite ice cream?

    c) How many students chose each flavor?

    d) How many students were in the group altogether?

16) In a box of 50 marbles, there are 15 red marbles, 20 white marbles, and the remaining marbles are blue.

    a) Write the ratio of each color marble to the total number in the box.

    b) Which of these ratios is/are equal to $\frac{2}{5}$?

17) Write each as an hourly rate.

    a) 180 miles driven in 3 hours

    b) $75 earned in 5 hours

18) Use order of operations to evaluate.

    a) $12 - 2^3 + 5 \times 6$

    b) $8 + 18 \div 2 \times 5$

19) Add:

    a) $\frac{2}{3} + \frac{7}{12}$

    b) $3\frac{1}{2} + 5\frac{3}{4}$

    c) $-18 + 7$

    d) $-12 + -6$

    e) $12.3 + 56.8 + 3.45$

20) Subtract:
    a) $\frac{3}{5} - \frac{1}{3}$
    b) $4\frac{2}{5} - 3\frac{1}{3}$
    c) $-12 - (-7)$
    d) $18 - (-8)$
    e) $8.7 - 4.28$

21) Multiply:
    a) $\frac{1}{2} \times \frac{4}{5}$
    b) $2\frac{1}{3} \times 5\frac{1}{6}$
    c) -12×-4
    d) -11×5
    e) 7.8×8.2

22) Divide:
    a) $\frac{3}{4} \div \frac{1}{2}$
    b) $2\frac{1}{5} \div \frac{2}{5}$
    c) $-18 \div 6$
    d) $-25 \div -5$
    e) $6.552 \div 4.2$

23) Find the GCF of 25 and 45.

24) Find the LCM of 18 and 12.

25) A marble is chosen at random from a jar that contains 2 red marbles, 2 blue marbles, and 4 yellow marbles. Find the probability of each:
    a) the marble is red
    b) the marble is blue
    c) The marble is yellow
    d) the marble is purple

26) What is the complement of a 40-degree angle?

27) What is the supplement of 75-degree angle?

28) Find x, then find all the angles.

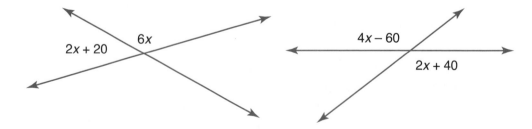

29) Find the solution set of the system of equations.

$x + y = 11$
$3x - 2y = 8$

$2x = 3y$
$4x - 3y = 12$

30) Tell whether the sequence is arithmetic, geometric, or neither.
a) 1,   4,   16,   64,   256
b) 6,   4,   2,   0,   −2
c) 0,   8,   16,   24,   32
d) 1,   6,   10,   13,   15

31) Match the example to the property
1. $4 \times 3 = 3 \times 4$
2. $5 + 0 = 5$
3. $6(x - 6) = 6x - 36$
4. $6 \times (2 \times 5) = (6 \times 2) \times 5$
5. $4 \times \frac{1}{4} = 1$
6. $x + 8 = 8 + x$
7. $7 \times 0 = 0$
8. $7 + -7 = 0$

a) Commutative property of addition
b) Commutative property of multiplication
c) Associative property of multiplicatoin
d) Additive identity
e) Multiplication by zero
f) Additive inverse
g) Multiplicative inverse
h) Distributive property

32) Find the volume and surface area.

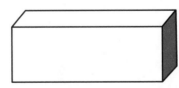

$r = 4$ in $l = 7.2$ m $w = 3$ m
$h = 7$ in $h = 4.3$ m

33) Find the perimeter and area.

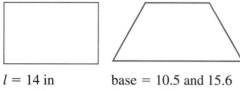

$l = 14$ in   base = 10.5 and 15.6
$w = 6$ in   sides = 4.2
                    height = 3.3

34) Find the circumference and area.

Radius = 9 in

35) Create a scatter plot.

Hours worked	2	3	5	7	8	12	14	16	17
Money earned	12	16	25	38	42	65	70	82	86

Answers to Chapter Reviews

1)

	Counting Number	Whole Number	Integer	Rational	Irrational	Real
5	x	x	x	x		x
0		x	x	x		x
−2			x	x		x
$\frac{1}{2}$				x		x
π					x	x
$\sqrt{6}$					x	x
$\sqrt{16}$	x	x	x	x	x	x

2)
1. B
2. D
3. H
4. C
5. G
6. A
7. E
8. F

3) 5, 1.5, 216

4) Incorrect
The division should be done first because it comes first.
$50 - 18 \div 2 \times 3$
$50 - 9 \times 3$
$50 - 27$
23
The answer is 23.

Chapter 2

1) $1\frac{3}{8}$

2) $10\frac{5}{12}$

3) $\frac{1}{8}$

4) $1\frac{1}{8}$

5) $1\frac{3}{4}$

6) $\frac{1}{4}$

7) 4

8) $16\frac{1}{2}$

9) $\frac{2}{9}$

10) $4\frac{1}{4}$

11) Day 8

12) $2\frac{5}{6}$ hours or 2 hr 50 min

Chapter 3

1) $5.94

2) $11.05

3) $17.70

4) $12.35

5) 26,400.62

6) 12.035   12.35   12.53   15.423

Chapter 4

1. a) -24
   b) -28
   c) -28
   d) 18

2. a) 3
   b) -12

c) -3
d) -5
e) -5

3. -6 degrees (6 hours $\times -6$ degrees $= -36$; $30 - 36 = -6$)

Chapter 5

1) $-18xyz$

2) $12x^2y$

3) $3ab^2$

4) $9a^3b^4$

5) $-2x^2 + 9x - 8$

6) $19x^2 - 10y + 6$

7) They are not like terms because they do not have the same variables.

8) $4d - 6$; 26

9) $d + 8$; 16

10) $d - 5; 3$

11) $\frac{d}{4} - 5; -3$

12) $7 + -5d; -33$

Chapter 6

1) a. $\frac{7}{6}$

   b. $\frac{2}{1}$

   c. $\frac{1}{4}$

   d. $\frac{5}{21}$

2) Yes, $\frac{6}{8}$ reduces to $\frac{3}{4}$; they are equivalent

3) 7.5 km

4) 5 hours

5) 388.5 grams

6) No, $\frac{3}{4}$ does not equal $\frac{5}{6}$

Chapter 7

1) 60%

2) 4.1%

3) $64.80

4) $1,125

5) $8,400

Chapter 8

1) a) 8

   b) $\frac{1}{16}$

   c) 1

   d) 15

   e) 4

   f) 48

   g) 14

2) a) 2^9

   b) $12^3 \cdot 3^3$

3) a) 2^{17}

   b) $2^9 \cdot 3^7$

   c) $3^7 \cdot 5^{-2}$

   d) 2^{-5}

   e) $2^{-1} \cdot 3^{-4}$

   f) $3^7 \cdot 5^{-2}$

4) a) 1.24×10^{10}

   b) 8.37×10^{-6}

5) a) 0.0000000036

   b) 5,400,000

6) a) 8

   b) 15

7) a) $2 \cdot 3^3$

   b) $5 \cdot 13$

8) a) 24

   b) 75

9) 1,130

Chapter 9

1) $-4 + 4 = 0$, not -8; $x = -3$

2) Subtract 3, not add; divide by -6; $a = -2$

3) $x = 2$

4) $b = -11$

5) $x = 2$

6) $x = -1$

7) $p = -10$

8) $y = -3$

9) $y = -4$

10) $p = 0$

11) 11 by 13

12) 13.3%

13) 40%

14) 6 lbs.

15) $1\frac{11}{60}$ or $\frac{71}{60}$

16) $2^3 \times 7$

17) 11, 22, 33, 44, 55, 66, 77

18) $.49

19) 3

20) 11

Chapter 10

1) Choice 4

2) Choice 4

3)

![number line with closed circle at 8, arrow pointing left]

4)

![number line with closed circle at 5, arrow pointing right]

5)

![number line with open circle at 0, arrow pointing right]

6)

![number line with closed circle at 6, arrow pointing left]

7) $x < -4$

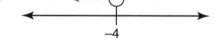

8) $x < 2$

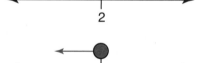

9) $x \leq -6$

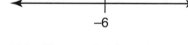

10) $3x + 50 \geq 100$ You need to have 17 or more sales.

11) $x \geq 57.14$ The amount in dollars needs to be $58.

12) $x \geq 6.3$ You need to sell at least 7 skirts.

Chapter 11

1) 30 degrees

2) 65 degrees

3) 72 and 108 degrees

4)

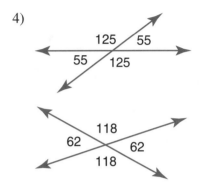

5)

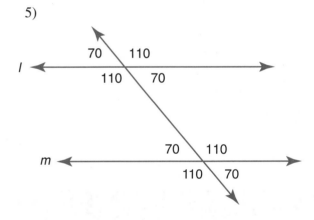

6) $2x + 45 + 3x + 5 = 180$
$$5x + 50 = 180$$
$$x = 26$$

7) Area = 3 cm²   Area: 12 in²
Perimeter = 7.4 cm   Perimeter = 16 in

8) Area = 12.25π ft²
Circumference: 7π ft

9) Volume: 351.9 in³   Volume: 122.0 cm³
Surface Area: 276.5 in²   Surface Area:
                             193.5 cm²

10) The new tank will hold 15,000 cm³. This is 15 liters. The new tank will hold 5 liters more.

Chapter 12

1)

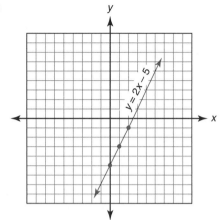

2)

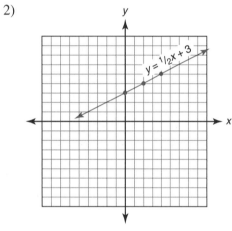

3) -2

4)   undefined

5) $-\frac{7}{6}$

6)

Equation	Slope	y-intercept
$y = 4x - 18$	4	-18
$y = -2x$	-2	0
$y = 2x - 7$	2	-7
$y = \frac{1}{3}x$	$\frac{1}{3}$	0
$y = 8x - 3$	8	-3

7)   domain (0,1,2)
     Range (3,5,7)
     Yes, it is a function.

8)   a)   linear
     b)   non-linear
     c)   linear
     d)   non-linear

Chapter 13

1)

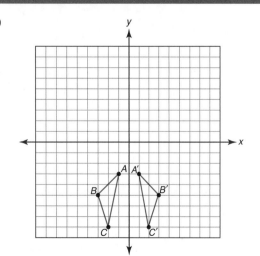

2)

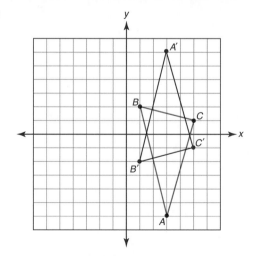

3)

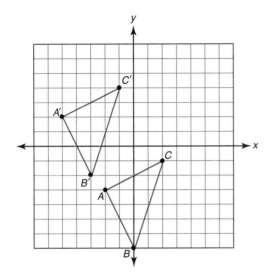

5)

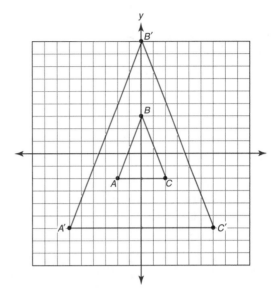

4)

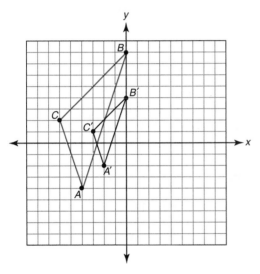

6)

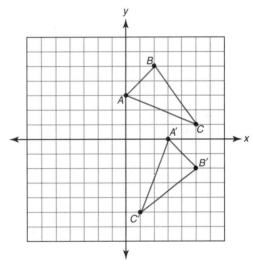

7)   A H I

8)   B C D E H I

Chapter 14

1)   Consistent; solution (6, 5)

2)   Dependent

3)   Consistent; solution (−3, 1)

4)   Inconsistent

5)   (3, 2)

6)   (−3, 2)

7)   (4, 1)

8)   Roses cost $3.00 each and carnations costs
$1.50 each.

Chapter 15

1) a) yes
   b) yes
   c) no

2) a) 13.7 ft.
   b) 72.8 ft.
   c) 79.1 in.
   d) 15 yd.

3) 12.2 miles

4) 31.2 feet

5) a) 6.4
   b) 5
   c) 3.2

Chapter 16

1) a) Permutation $P(20,2) = 380$
   b) Combination $C(36,4) = 58,905$
   c) Combination $C(12,5) = 792$
   d) Permutation $P(10,3) = 720$

2) $3 \times 4 = 12$

3) $3 \times 3 = 9$

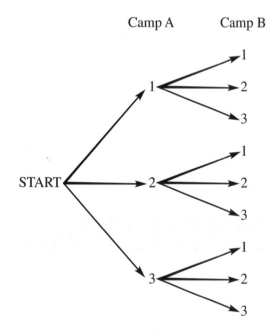

Camp A   Camp B

START

4) a) $\frac{1}{4}$
   b) $\frac{2}{4}$ or $\frac{1}{2}$
   c) $\frac{2}{4}$ or $\frac{1}{2}$
   d) $\frac{4}{4}$ or 1
   e) $\frac{0}{4}$ or 0

5) $\frac{1}{16}$

   1,1 1,2 1,3 1,4

   2,1 2,2, 2,3 2,4

   3,1 3,2 3,3 3,4

   4,1 4,2 4,3 4,4

6) a) $\frac{5}{15} \times \frac{4}{14} = \frac{20}{210} = \frac{2}{21}$
   b) $\frac{2}{15} \times \frac{1}{14} = \frac{2}{210} = \frac{1}{105}$
   c) $\frac{8}{15} \times \frac{2}{14} = \frac{16}{210} = \frac{8}{105}$

7) a) $\frac{5}{15} \times \frac{5}{15} = \frac{25}{225} = \frac{1}{9}$
   b) $\frac{2}{15} \times \frac{2}{15} = \frac{4}{225}$
   c) $\frac{8}{15} \times \frac{2}{15} = \frac{16}{225}$

Chapter 17

1) Her current average is 83.75. She needs to raise it by 1.25 points. She needs to get a 90 on her next test.

2) a) Mean = 7
       Median = 7.5
       Mode = 8
       Range = 3

   b) Mean = 319
       Median = 310
       Mode = no mode
       Range = 164

3) Four

4) Eight

5) Round-trip of commuters

0	3	4	6	9	
1	0	2	5	6	
2	0	6	9		
3	0	2	3	8	9
4	0	0	2	4	
Key 1	0 = 10				

   Box whisker
       Minimum = 3
       First Quartile = 11
       Median = 27.5
       Third Quartile = 38.5
       Maximum = 44

6)

Intervals	Frequency
11–20	2
21–30	3
31–40	5
41–50	4
51–60	6

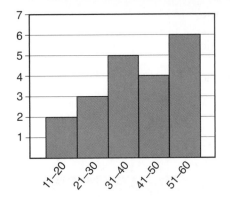

7)

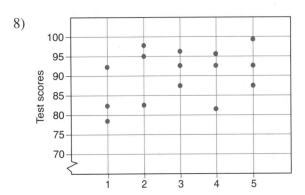

8)

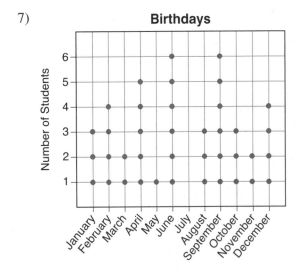

9) There is a positive correlation.

Chapter 18

a) geometric

b) arithmetic

c) arithmetic

d) neither

e) geometric

Cumulative Review

1) 60%

2) 4%

3) $64.80

4) Reflection over the y-axis
   A $(2, 2) \to A'(-2, 2)$
   B $(4, 2) \to B'(-4, 2)$
   C $(4, -4) \to C'(-4, -4)$

5) Reflection over the x-axis
   A $(2, 2) \to A'(2, -2)$
   B $(4, 2) \to B'(4, -2)$
   C $(4, -4) \to C'(4, 4)$

6) Translate 7 units right and 4 units up
   A $(-4, -1) \to A'(3, 3)$
   B $(-4, -3) \to B'(3, 1)$
   C $(-1, -1) \to C'(6, 3)$

7) Reflection over the line $x = 1$
   A $(4, -1) \to A'(-2, -1)$
   B $(4, -3) \to B'(-2, -3)$
   C $(-1, -1) \to C'(3, -1)$

8) Min $= 24$
   Q1 $= 42$
   Med $= 47.5$
   Q3 $= 56$
   Max $= 76$

9) a) $2 < n < 8$
   b) $8n - 2$
   c) $n - \frac{8}{2}$
   d) $2(n + 8)$
   e) $n - 8 < 2$
   f) $n + 2 = 8$
   g) $(n + 2) + 8$
   h) $n - 8 = 2$

10) a) $x = 4$
    b) $x = -4$
    c) $x = 8$
    d) $x = 4$
    e) $x = 15$
    f) $x = 23$
    g) $x = 7$
    h) $x = -10$
    i) $x = 3$

11) a) 2
    b) 5
    c) 3
    d) 1
    e) 6
    f) 4

12) a) $x \le -1$

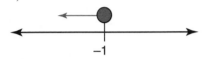

    b) $x > 7$

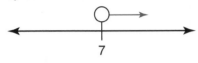

    c) $x < -4$

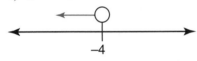

    d) $x < 3$

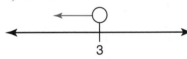

13) a) 6
    b) $\frac{1}{125}$
    c) 64

14) a) $c = 13$
    b) $b = 48$

15) a) chocolate
    b) strawberry
    c) chocolate 7
       vanilla 6
       strawberry 2
    d) 15 students

16) a) Red $\frac{15}{50}$ or $\frac{3}{10}$
       White $\frac{20}{50}$ or $\frac{2}{5}$
       Blue $\frac{15}{50}$ or $\frac{3}{10}$
    b) White

17) a) 60 mph
    b) $15 per hour

18) a) 34
    b) 53

19) a) $1\frac{1}{4}$
    b) $9\frac{1}{4}$
    c) -11
    d) -18
    e) 72.55

20) a) $\frac{4}{15}$
    b) $1\frac{1}{15}$
    c) -5
    d) 26
    e) 4.42

21) a) $\frac{2}{5}$
    b) $12\frac{1}{18}$
    c) 48
    d) -55
    e) 63.96

22) a) $1\frac{1}{2}$
    b) $5\frac{1}{2}$
    c) -3
    d) 5
    e) 1.56

23) 5

24) 36

25) a) $\frac{2}{8}$ or $\frac{1}{4}$

   b) $\frac{2}{8}$ or $\frac{1}{4}$

   c) $\frac{4}{8}$ or $\frac{1}{2}$

   d) 0

26) 50 degrees

27) 105 degrees

28) $x = 20$ the angles are 60 and 120
   $x = 50$ the angles are 140 and 40

29) (6, 5)
   (6, 4)

30) a) geometric
   b) arithmetic
   c) arithmetic
   d) neither

31) 1) b
   2) d
   3) h
   4) c
   5) g
   6) a
   7) e
   8) f

32) volume $= 112\pi$ or 352 in³
   volume $= 92.88$ m³
   surface area $= 88\pi$ in² or 276 in²
   surface area $= 130.92$ m²

33) $P = 40$ in $A = 84$ in²
   $P = 34.5$ $A = 43.065$

34) $C = 18\pi$ $A = 81\pi$

35)

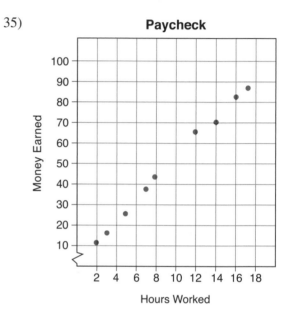

INDEX

A

Acute triangle, 135
Addition property of zero, 4
Additive identity, 4
Additive inverse, **4**, 200–201, 205
Algebraic equations
   angles, 126, 132–134
   binomial, 44
   common factor, 89
   corresponding, 131
   functions, 167–168
   geometry, 128–130, 153–155,
      175–179
   inequality, 114
   patterns, 263
   rule of four, 168–169
   simplifying, 46
   solving, 109–111
   word problems, 204
Alternate interior angles, 131
Angles, **124**, 125
   algebraic equations, 126,
      132–134
   bisector, 125
   classify a triangle, 135–136
   corresponding, 131
   one-step equations, 100–102, 125
   pairs, 124
   solving unknown, 128–130
   types, 124
   vertical, 131
Area
   formulas, 138–142
   Pythagorean Theorem, 210
   surface area formula, 146–150
Arithmetic sequences, 265–266
Associative property, **4**
Axis(es), 152

B

Bar graph, **234**, 235, 242
Base, **84**
   in exponents, 85
   as a factor, 85
   known as faces, 143
   law of exponents, 91–94
   pyramids and prisms, 143
   ten number system, 23
   three-dimensional figures, 143
Binomial, 44
Box-Whisker, 250–252

C

Cartesian plane, 152, 255
Center, 137, 144
Central tendency, 240–241
Circle graph, **235**, 244
Circles, **137–138**
   area, 140
   circumference, 137, 141
   diameter, 141
   related to a cone, 144
Circumference, 137, 141
Closed-plane figure, 135–138
Coefficients, 44
Combination, 230
Commission, 74
Common difference, geometry, 265
Common factor, 89
Common multiple, 89
Common ratio, 266–267
Commutative property, 4, 7, 39, 213,
   273
Comparison
   in bar graphs, 242
   ratios, 52–53
   y vs. x coordinate for slope,
      158–159
Complementary angles, 124
Composite number, 87
Compound event, 220
Computation rate, 54
Cone, **144**, 148, 149
Congruent, 52, 58, 127, 143, 144, 179,
   180, 183, 187
Congruent angles, **124**, 125, 128, 130,
   131, 136–137
Congruent sides, 136, 137
Constant, 45, 171
Constants, **44**
Converse of the Pythagorean Theorem,
   211
Coordinate geometry, 153–155
Coordinate plane, 152
Coordinates, 152
Correlation, 255
Corresponding angles, 131
Counting numbers, 2
Counting principle, 221
Cross-multiplying, 56
Cross-products, 57, 59, 61
Cube numbers, 264
Cylinder, **144**, 147, 148

D

Decimal numbers, 2
Decimals, **23–24**
   add and subtract, 26–28
   base ten number system, 23
   comparison of, 23–25
   dividing, 30–31
   multiplying, 28–30
Denominators, 10
Dependent events, 227
Diameter, 137
Dilation, **177**, **188**, 189
Discount, 75
Distributive property, **5**, 106–107,
   202–204
Dividing fractions, 19–20
Division, 52–53
Divisor, 30
Domain of a function, 164
Dot plots, 246–247
Double-bar/line graph, 235

E

Edge, 143
Equality
   one-step equation, 100–102
   substitution property, 48
   two-step equation, 102–104,
      106–107
Equations, **100**
   equality, 100
   first/second degree, 172
   graphing from a table of values, 155
   inequality, 114
   like terms, 104
   linear vs non-linear, 171–173,
      195–197
   multi-step equations, 107–109
   one-step equation, 100–102, 125
   percent, 70–71
   probability, 224
   or proportions, 70, 71
   Pythagorean Theorem, 210
   slope-intercept form, 160, 197
   solving algebraically, 109–111
   two-step equation, 102–104,
      106–107
Equilateral triangle, 136
Equivalent fractions, 10–12
Estimation, 28
Evaluate, **48–49**

Event, 220
Exponential form of a number, 94–95
Exponents, **5**, **84–86**, 91–94
Expressions, **45**
    evaluating, 48–49, 96
    with exponents, 96
    and order of operation, 6
    percent, 68
    simplifying, 46–47
    writing, 45–46

F
Face, 143–144
Factor, **84**, 90
Fibonacci numbers, 264
Figures
    closed-plane, 135–138
    similar, 58–60
    three-dimensional figures, 143–146
    two-dimensional, 135–138
"Find the elevator," 153
First degree equation, 171
Focal point, 185
Formulas
    area, 138–140
    circumference of a circle, 141
    context clues, 45
    diameter of a circle, 141
    percent error, 80
    perimeter, 140
    surface area, 146–150
    using symbols, 44–45
    volume, 147–149
    and writing expressions, 45–46
Fractions, **2**, **10**
    adding, 12–15
    dividing, 19–20
    equivalent, 10–12
    exponents, 85
    multiplying, 18–19
    proportion, 55–56
    related to decimals, 24
    subtracting, 15–17
    word problems, 20–21
Frequency table, 252
Function, 155, **164**–168, 188

G
GCF, *see* Greatest common factor
Geometric sequences, 266
Graphing
    box-whisker, 250–252
    dot plots, 246–247
    from an equation, 155
    function, 168–169
    interpretation, 241

linear vs non-linear, 171–173
    pictograph, 235
    picturegraph, 243
    pie graph, 235
    with Pythagorean Theorem, 215–216
    stem-leaf plot, 248–250
    systems of linear equation, 197–198
    table of values, 155–156
    types of graphs, 234–240
Gratuity, 76
Greatest common factor (GCF), 89

H
Histogram, **252**–255
Hypotenuse, 210

I
Identical sets, 131
Image, **178**, 180, 182–183, 187
Improper fraction, 10
Income, 73
Independent events, 226
Independent variable, 155, 164
Inequality, **114**, 116–121
Integers, **33**–38
Interest, 75
Inverse operations, 101, 102
Irrational numbers, 2
Isosceles triangle, 136

L
Lateral faces, 143
Law of exponents, 91–94
LCD, *see* Lowest common
        denominator
LCM, *see* Least common multiple
Least common multiple (LCM), **10**, **89**
    adding fractions, 14
    examples, 90
    mixed numbers, 14
    subtracting fractions, 16
    in systems of linear equation, 200
Leg(s) of a right triangle, 210
Line, 127, 131
Line graph, 235, 242
Line of symmetry, 179
Linear pair, 127, 132
Lowest common denominator (LCD),
        10, 13

M
Map scale, 52
Mathematical relationship, 168–169
Mean, 240
Means of a proportion, 52
Measures of central tendency, 240–241

Median, 240, 250
Mode, 241
Money, 24, 29
Monomial, 44
Multiplicative identity, 4, 11
Multiplicative inverse, 5
Multiplicative property of one/zero, 4
Multiplying, 18–19, 56
Multi-step equations, 107–109

N
Nanometer, 94
Net, 144
Non-linear line, 172
Notations, 52–53, 116
Number line, 115–116
Numbers, **1**
    decimals, 23–25
    exponential form, 94–95
    inequality symbols, 117
    integers, 33–35
    irrational, 2
    patterns, 263–264
    rational, 2
    real, 2, 4
    square root, 5
    standard form, 94
Numerator, 10
Numeric pattern, 263

O
Obtuse triangle, 136
One-step equation, 100–102, 125
Order of operations, 2–7
Ordered pairs, 155, 195, 220–221
Origin, 185

P
Parabola, 172
Parallel lines, 131
Parallelogram, 136, 139
Patterns, 263–264
Percent, **66**
    calculation of, 67–70
    as commission of sales, 74–75
    decrease/increase, 78
    discount and sales price, 75–76
    error, 78, 80–81
    as gratuity, 76–77
    as interest rate, 75
    ratios, 78
    as sales tax, 72–73
    as withholding tax, 73–74
Perimeter, 140
Permutation, 229–230
Pictograph, 235

Picture graph, 243
Pie graph, 235
Place value, 24
Placement of commas, 45
Point, 153, 158
Point symmetry, 192
Polygons, 135–136, 138–140, 143
Polynomial, 44
Population, 234
Power of ten, **24**, 30
Pre-image, **178**, 180, 182–183, 187
Preserved image, 178
Prime factorization, 87–88
Prime number, 87
Prism, **143**, 146, 147
Probability, **223**, 224–228
Product, 84
Proper fraction, 10
Properties of real numbers, 4
Proportion, **52**
    as fractions, 55–56
    percent, 69
    scale drawings, 60
    sides of similar figures, 58
    solving, 57–58
    word problems, 62–63
Pyramid, 143–144
Pythagorean Theorem, **210**, 211–216
Pythagorean triples, 210

Q

Quadrant, 152
Quadrilateral types, 136–137

R

Radius, 137
Random sample, 234
Range of a data set, 241
Range of a function, 164
Rate, **52**, 54, 160
Ratio, **52–53**
    of change, 158
    of change in percentages, 78
    common as related to sequences,
        266–267
    division, 52–53
    as a percent, 66
    probability, 223–224
    slope, 161
    word problems, 62–63
Rational numbers, 2
Real numbers, 2, 4
Reciprocal, 10
Rectangle, 136
Reflection, 176, **179**, 180
Regrouping of decimals, 27

Relative error, 80
Rhombus, 136
Right triangle, 135, 210–211
Rotation, 176, 185, 187
Rotational symmetry, 185
Rule of four, 168

S

Sale price, 75
Sample, **220**, 223, 234
Scale, **52**, 60
Scalene triangle, 136
Scatter plot, **255**, 256–258
Scientific notation, **94–96**
Second degree equation, 171
Sequences, **264–267**
Similar triangles, 52
Simple events, 224–227
Simple symmetry, 191
Slope, **158**
    calculation, 158–159
    graphing, 160–163
    intercept form, 160, 161, 197
    linear equation, 161, 170
    rate of change, 160
    systems of linear equation, 196
Slope-intercept form, 161
Solution, 100, 152, 196
Solution set, 115
Sphere, **144**, 148
Square, **136**, 139
Square numbers, 264
Square pyramid, 143
Square root of a number, 5
Squared number, 5
Standard form of a number, 94–95
Statistics, **234**, 234–236
Stem-leaf plot, 248–250
Substitute, **48**, 202–206
Substitution property, 48
Supplementary angles, 124,
    127, 130
Surface area, 146–147
Survey, 234
Symbols, **44–45**
    combination, 230
    in graphs, 235, 243
    inequality, 114
    permutation, 229
Symmetry, **191**, 192
Systems of linear equations
    solve by addition, 200–202
    solve by graphing, 197–199
    solve by substitution, 202–204
    types of, 195–197
    word problems, 204–206

T

Table of values, **155**, 171–173
Three-dimensional figures, 143–146,
    144
Transformation, **175**, 176–177
Transformational geometry, 175–179
Translation, **177**, **181**, 182–183
Transversal, 131
Trapezoid, 136
Tree diagram, **220–221**, 223
Triangle numbers, 263–264
Triangles
    area, 138
    classified by angles, 135–136
    classified by sides, 136
    Pythagorean Theorem, 210
    similar, 52, 58–60
Triangular pyramid, 143
Trinomial, 44
Two-dimensional figures, 135–138
Two-step equation, 102–104,
    106–107

U

Unit fraction, 10
Unit price, 52
Unlike denominators, 10

V

Variable, **44**, 100–102, 170–172
Vertex, 143
Vertical angles, 127
Vertical line test, 165
Volume formulas, 147–148

W

Whole numbers, 2, 24
Word problems
    fractions, 20–21
    inequalities, 119–121
    proportions and ratios, 62–63
    Pythagorean Theorem, 213
    scale drawings, 61
    in systems of linear equation,
        204–206

X

X-axis, **152**, 153, 179
X-intercept, **152**, 196

Y

Y-axis, **152**, 153, 179
Y-intercept, **152**, 196

Z

Zero pairs, 36, 38

MOVE TO THE HEAD OF YOUR CLASS
THE EASY WAY!

Barron's presents **THE E-Z SERIES** (formerly THE EASY WAY SERIES)—specially prepared by top educators, it maximizes effective learning while minimizing the time and effort it takes to raise your grades, brush up on the basics, and build your confidence. Comprehensive and full of clear review examples, **THE E-Z SERIES** is your best bet for better grades, quickly!

Available at your local book store
or visit **www.barronseduc.com**

Barron's Educational Series, Inc.
250 Wireless Blvd.
Hauppauge, NY 11788
Order toll-free: 1-800-645-3476
Order by fax: 1-631-434-3217

In Canada:
Georgetown Book Warehouse
34 Armstrong Ave.
Georgetown, Ontario L7G 4R9
Canadian orders: 1-800-247-7160
Order by fax: 1-800-887-1594

(#45) R7/12

Prices subject to change without notice.

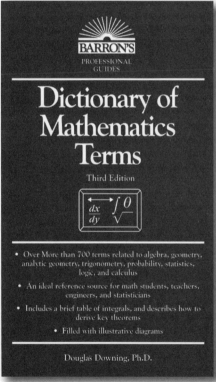

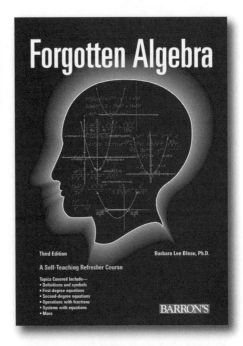
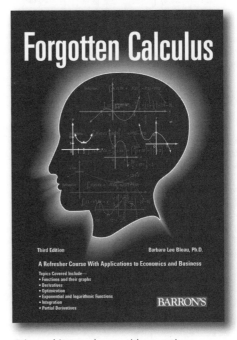